Texts and
Monographs
in Physics

Andrzej Derdzinski

Geometry of the Standard Model of Elementary Particles

 Springer-Verlag Berlin Heidelberg GmbH

Professor Andrzej Derdzinski
Dept. of Mathematics
Ohio State University
Columbus, OH 43210, USA

Editors

Wolf Beiglböck

Institut für Angewandte Mathematik
Universität Heidelberg
Im Neuenheimer Feld 294
D-6900 Heidelberg 1, FRG

Tullio Regge

Istituto di Fisica Teórica
Università di Torino
C. so M. d'Azeglio, 46
I-10125 Torino, Italy

Robert P. Geroch

Enrico Fermi Institute
University of Chicago
5640 Ellis Ave.
Chicago, IL, 60637, USA

Walter Thirring

Institut für Theoretische Physik
der Universität Wien
Boltzmanngasse 5
A-1090 Wien, Austria

Elliott H. Lieb

Department of Physics
Joseph Henry Laboratories
Princeton University
Princeton, NJ 08540, USA

ISBN 978-3-642-50312-2   ISBN 978-3-642-50310-8 (eBook)
DOI 10.1007/978-3-642-50310-8

Library of Congress Cataloging-in-Publication Data
Geometry of the standard model of elementary particles / Andrzej Derdzinski. p.cm. – (Texts
and monograms in physics) Includes bibliographical references (p.) and index.

1. Standard model (Nuclear physics) 2. Electroweak interactions. I. Title. II. Series.

Originally published by Springer-Verlag Berlin Heidelberg New York in 1992.

Data conversion by Type 2000, Mill Valley, USA;

55/3020-5 4 3 2 1 0 – Printed on acid-free paper

Preface

The standard model is the currently accepted theory of particles and interactions, describing them in a manner that is both coherent and consistent with experimental data. This book deals only with the *classical part* of the standard model, obtained by ignoring its aspects related to field quantization. Despite its shortcomings as a description of nature, such a classical approach offers a simple yet far-reaching insight into the structure of particle theories (see end of §6.7).

The reader need not be familiar with theoretical physics, but is expected to know the basic facts about Riemannian manifolds and connections in vector bundles, as presented, e.g., in the appropriate introductory chapters of [Besse 1987], [Kobayashi and Nomizu 1963], or [Milnor and Stasheff 1974]. The approach used is based on geometric language (coordinate-free, whenever possible), and should be easily accessible to mathematicians, as well as physics students with some background in geometry.

The chapters and sections marked by asterisks are considered optional and may be skipped, leaving an essentially self-contained, 67-page-long core of the text. It covers the most important topics (briefly summarized in [Derdzinski 1991a]), such as the Yang-Mills description of interactions, the quark and electroweak models, and the SU(5) grand unification. The optional chapters deal with further material, including the geometry of (iso)spin, particle invariants, unitary symmetry, the Cabibbo angle, Lagrangian densities, and the Higgs mechanism.

The purpose of this book is to fill what I perceive as a gap in the available literature. There exist many excellent accounts of these and related topics, both in the form of physics books and mathematical texts (see the reference list). However, while the former usually require much training in physics, and describe geometric constructions in terms of coordinates, the latter's emphasis is on mathematics and generality. My objective has been to present the classical version of the theory believed to reflect physical reality, by means of a consistently geometric exposition that avoids extensive generalizations, and assumes no prerequisite knowledge of physics.

Columbus, February 1992   *Andrzej Derdzinski*

Acknowledgements

The present book evolved from talks I gave in mathematics seminars at the Ohio State University in the spring and summer of 1989. I would like to express my gratitude to Alexander Dynin, Daniel Dynin and Robert Stanton, who have helped me with discussion and advice, and to Elliott Lieb, who invited me to contribute the book to this series.

I also wish to thank Janice McDonald, who did an excellent job of typing this manuscript.

During the preparation of this book I was supported in part by the National Science Foundation under Grant No. DMS-8601282.

Andrzej Derdzinski

Contents

Part I

Particle Models

1 Particles and Vector Bundles

This chapter deals with some specific vector bundles over the spacetime that serve as models of elementary particles. We restrict our consideration to *particles of matter* (as opposed to particles mediating interactions) which are *free*, i.e., not subject to interactions. The choice of the bundles to represent them is justified in Chapter 3. Analogous models of interaction carriers and interacting particles are described in Chapters 5 and 6.

Also included here is a brief presentation of particles and interactions encountered in nature.

1.1 Spacetimes

A *spacetime* is a 4-dimensional pseudo-Riemannian manifold $(\mathcal{M}, g)$ of the Lorentz signature $-+++$, endowed with a fixed *time orientation*, i.e., a continuous selection $\mathcal{M} \ni x \mapsto B_x$ of one ("future") connected component B_x of the set $\{v \in T_x\mathcal{M} : g(v,v) < 0\}$ of *timelike* vectors at x. (In particular, a spacetime has to be *time-orientable*.)

The simplest example is the *Minkowski spacetime M*, for which the underlying manifold is an affine 4-space and the metric g is translation-invariant. It constitutes the geometric model of Einstein's *special theory of relativity*, which accounts, among other things, for the experimental fact that the speed c of light in the vacuum appears the same to all inertial observers (Remark 4.3.1). See also §§4.2, 4.3 and Remarks 6.9.5, 6.9.6.

Nonflat spacetimes $(\mathcal{M}, g)$ are used in Einstein's *general relativity*, where the curvature of g represents the part of gravity effects that cannot be removed by acceleration, i.e., the tidal forces.

References: Sachs and Wu 1977; Dixon 1978; Besse 1987; Emch 1984; Derdzinski 1991b.

1.2 Elementary Particles

By a *particle* we mean an elementary-particle species, such as the electron, denoted e, the proton p, neutron n, photon γ, etc. In the semiclassical approach we are using (without field quantization), each particle is represented by, or "lives" in, a bundle η with some additional structure (geometry) over a fixed spacetime $(\mathcal{M}, g)$. Those geometries of η actually used in particle physics are quite specific (see §§1.16, 1.17, 5.4). In particular, η is usually a real or complex *vector* or *affine* bundle and its geometry includes some *field equations*, which form a system of partial differential equations imposed on sections of η. The sections ψ of η then may be called the *semiclassical states* of the corresponding particle.

Let us now consider a spacetime manifold $\mathcal{M}$ diffeomorphic to $\mathbb{R} \times \mathbb{M}$ for a 3-manifold $\mathbb{M}$ (e.g., the Minkowski spacetime), with fixed identifications $\mathcal{M} = \mathbb{R} \times \mathbb{M}$, $\eta = \mathrm{pr}^*\zeta$, where ζ is a vector bundle over $\mathbb{M}$ and $\mathrm{pr} : \mathcal{M} \to \mathbb{M}$ is the projection. Sections ψ of η then become curves $t \mapsto \phi(t, \cdot)$ of sections of ζ parametrized by the "time" $t \in \mathbb{R}$, and the following two natural (but in part conflicting) interpretations of the semiclassical states ψ are possible:

(a) One may think of sections ϕ of ζ as possible configurations of an elastic medium filling the space $\mathbb{M}$, whose equilibrium position corresponds to the zero section. Our ψ then are disturbances of this medium ("waves") that evolve with time (i.e., are propagated) according to the field equations.

(b) The ψ can be regarded as evolving *quantum states* in ζ (i.e., L^2 sections of ζ, varying with time), the field equations being an analogue of Schrödinger's equation. In contrast with (a), multiplying ψ by nonzero numbers now leaves its physical properties unchanged, while the zero section of η or ζ has no physical meaning at all. See §2.5.

The fibres η_x of η, $x \in \mathcal{M}$, can be expected to represent the *internal structure* of the corresponding particle. In fact, this is obvious when $\mathcal{M} = \mathbb{R} \times \mathbb{M}$ and $x = (t, \mathbf{x})$, as η_x then is the only part of the structure left after one has singled out the time t and location $\mathbf{x}$ and ignored everything else. Since the fibres of η must have a positive dimension in view of (a) or (b) above, the "simplest" possible fibres are represented by the product line bundles $\eta = \mathcal{M} \times \mathbb{C}$ and $\mathcal{M} \times \mathbb{R}$. Any particle species living in $\mathcal{M} \times \mathbb{C}$ or $\mathcal{M} \times \mathbb{R}$ is therefore regarded as *pointlike*, i.e., *structureless*.

1.3 Spinors

This section provides an ad hoc description of spinor bundles over space-times. An alternative definition, phrased in terms of spin structures, can be found in §6.3.

A *Weyl pair* (S, Ω) consists of a complex 2-dimensional vector space S and a nonzero skew-symmetric bilinear form $\Omega : S \times S \to \mathbb{C}$. Regarding S as a real 4-space endowed with $I \in \mathrm{End}_{\mathbb{R}}S$, given by $I = \mathrm{i} \cdot \mathrm{Id}$, and with $\omega = 2\,\mathrm{Re}\,\Omega \in \Lambda^2_{\mathbb{R}}S^* \smallsetminus \{0\}$, we can replace (S, Ω) by the *Weyl triple* (S, I, ω), characterized by $I^2 = -\mathrm{Id}_S$ and $I^*\omega = -\omega$. More general objects are *Majorana triples* $(S, \pm I, \omega)$, which differ from Weyl triples by having I only defined up to a sign, so that they may be thought of as Weyl triples (or pairs) "up to complex conjugation".

The *associated vector space* of a Weyl pair (S, Ω) is the real vector space $\mathcal{T}$ of all antilinear self-maps B of S for which the sesquilinear form $S \times S \ni (\psi, \phi) \mapsto \Omega(\psi, B\phi) \in \mathbb{C}$ is Hermitian. $\mathcal{T}$ is 4-dimensional and carries a natural Lorentz inner product $\langle,\rangle$ with $BB' + B'B = 2\langle B, B'\rangle \cdot \mathrm{Id}_S$ for $B, B' \in \mathcal{T}$, as well as a natural time orientation and orientation. For Majorana triples, $\mathcal{T}$ is still well-defined and so are $\langle,\rangle$ and the time orientation of $\mathcal{T}$, but *not* its orientation (in fact, there is a natural bijection between $\{I, -I\}$ and the orientations of $\mathcal{T}$).

A *Weyl spinor bundle* over an *oriented* spacetime $(\mathcal{M}, g)$ consists of a complex vector bundle $\sigma^{\mathrm{W}} \to \mathcal{M}$ along with C^∞ sections Ω and γ of $\Lambda^2(\sigma^{\mathrm{W}})^*$ and $\mathrm{Hom}(T^*\mathcal{M}, \mathrm{End}_{\mathbb{R}}\sigma^{\mathrm{W}})$, respectively, such that at each $x \in \mathcal{M}$, $(\sigma^{\mathrm{W}}_x, \Omega_x)$ is a Weyl pair and γ_x is a time-orientation and orientation preserving linear Lorentz isometry of $T^*_x\mathcal{M}$ onto the corresponding associated vector space $\mathcal{T}_x$. Similarly, a *Majorana spinor bundle* over a spacetime $(\mathcal{M}, g)$ (which may even be non-orientable) is a real vector bundle $\sigma^{\mathrm{M}} \to \mathcal{M}$ along with C^∞ sections $\pm I, \omega, \gamma$ of the appropriate bundles, forming at each $x \in \mathcal{M}$ a Majorana triple $(\sigma^{\mathrm{M}}_x, \pm I_x, \omega_x)$ with a time-orientation preserving linear Lorentz isometry $\gamma_x : T^*_x\mathcal{M} \to \mathcal{T}_x$. Thus, $\pm I$ is double-valued, with smooth *local* branches.

References: Lawson and Michelsohn 1989; Penrose and Rindler 1984–1986; Derdzinski 1991b.

1.4 Interactions

Experimental evidence indicates that in nature there exist the following kinds of interactions: *strong, electromagnetic, weak* and *gravitational*, ordered here by decreasing *strength* on the microscopic level, i.e., the probability that in the given circumstances the interaction will take place at all.

Gravitational forces are far too weak to cause measureable microworld effects. Therefore, in models to be discussed, gravity will only be a "background", represented by the geometry of the spacetime $(\mathcal{M}, g)$ in question (§1.1).

Electromagnetic interactions take place between particles carrying electric charges or magnetic moments. In particular, it is the electric forces involving nuclei and electrons that hold atoms together. Interatomic and intermolecular forces arise, within a specific distance range, as *residual effects* of the electric attraction/repulsion forces between constituents of different atoms. Thus, apart from gravity, most forces encountered at the macroscopic scale are of electromagnetic origin (here belong elasticity and strength of materials, viscosity, adhesion, etc.).

Strong forces are in particular responsible for forming atomic nuclei from nucleons (protons and neutrons), often against electromagnetic repulsion. According to the quark model (§6.2), supported by plentiful evidence, nucleons consist of certain peculiar particles known as *quarks*, which exert onto one another extremely strong *interquark forces*. The strong forces affecting nucleons are merely residual effects of these. Thus, there is a natural analogy between the nuclear and intermolecular forces, as opposed to the "pure" strong and electromagnetic interactions.

Weak interactions manifest themselves mainly through various decay processes. The most notorious of them is the *radioactive β decay*, during which a neutron is turned into three particles: a proton, an electron, and an electronic antineutrino ($\text{n} \rightarrow \text{p} + \text{e}^- + \bar{\nu}_e$).

References: Sudbery 1986; Ryder 1986; Leite Lopes 1981.

1.5 Matter Fields Versus Interaction Carriers

Various sorts of evidence indicate that interactions among particles proceed with the aid of ("are mediated by") certain special kinds of particles, known as *intermediate bosons*, or *interaction field quanta*. These are the *photons* for the electromagnetic interaction (a single species), 8 kinds of *gluons* for the strong interaction, and the *weak (intermediate) bosons* W^+, W^- and Z^0. Superscripts $^0, ^+, ^-, ^{++}$, etc., if present, stand for the electric charge measured in proton charge units. Thus, $\text{n} = \text{n}^0$, $\text{p} = \text{p}^+$, $\text{e} = \text{e}^-$, $\bar{\nu}_e = \bar{\nu}_e^0$. The photon $\gamma = \gamma^0$, and all gluons, are electrically neutral.

Similarly, gravitational forces are believed to be carried by special particles, called *gravitons*.

"Ordinary" particles which are not the interaction carriers just listed, are referred to as matter particles or *matter fields*. On the formal level, the distinction between both categories is considerable. While matter fields live in some specific vector bundles over the given spacetime $(\mathcal{M}, g)$ (§1.16, §1.17), the semiclassical states (§1.2) of interaction carriers turn out to be

most naturally represented by *gauge fields,* i.e., connections in certain fixed
"auxiliary" vector bundles (with additional structure) over $\mathcal{M}$. See §5.4.
However, under suitable circumstances, descriptions of intermediate bosons
in terms of sections of *vector* bundles may also be appropriate, cf. §5.8.

According to commonly used terminology, matter particles are divided
into *hadrons* and *leptons,* depending on whether or not they participate
in strong interactions. So far, experimenters have discovered 12 species of
leptons (including electrons, muons and neutrinos) and more than 200 kinds
of hadrons (the exact number cannot be determined, as these discoveries
have varying degrees of certainty). All known leptons and the lightest of
known hadrons are listed in §§1.14, 1.15.

References: Sudbery 1986; Ryder 1986; Leite Lopes 1981.

1.6 Antiparticles

Most particles live in *complex* vector bundles η over the given spacetime
$(\mathcal{M}, g)$. Let $\bar{\eta}$ be the *conjugate* bundle of such a bundle η. Thus, the fibre $\bar{\eta}_x$
of $\bar{\eta}$ over any $x \in \mathcal{M}$ has the same underlying real vector space structure as
η_x, while the multiplication by $z \in \mathbb{C}$ in $\bar{\eta}_x$ equals the multiplication by $\bar{z}$ in
η_x. With naturally defined conjugates of all objects forming the geometry of
η, the bundle $\bar{\eta}$ seems as good a candidate to host a particle, as η was. (See
(iv) of §1.19.) The particle we may thus expect to live in $\bar{\eta}$ is called the *antiparticle* of the particle species represented by η. One can easily deduce how
this procedure of *antiparticle formation* affects various numerical invariants
associated with particle species. All "reasonably" defined invariants turn
out to either remain unchanged, or simply undergo a change of sign, which
is clear as the formation of antiparticles is involutive. For instance, a particle and its antiparticle have equal masses (since the mass appears in the
field equations, which do not change under conjugation, as they are *real* in
all relevant cases), but their electric charges are opposite (see §4.14).

Experimental data indicate that *every* particle species has its antiparticle. The first example discovered in nature was the *positron* (antielectron)
e^+. Usually, with some traditional exceptions such as e^+, the antiparticle
is denoted by the symbol of the particle with a bar $\bar{\ }$: the antiproton $\bar{p}$,
antineutron $\bar{n}$, electronic antineutrino $\bar{\nu}_e$ (as opposed to the electronic neutrino ν_e). The antiparticle formation can be made applicable to interaction
carriers as well; for instance, the weak bosons W^+, W^- are each other's antiparticles ((ii)b) of §6.5). Some particle species, called *strictly neutral,* are
their own antiparticles, and then are naturally represented by *real* vector
bundles. See §1.7 for details and examples.

The identity transformation of the underlying real bundle gives rise to
the *canonical antiisomorphisms* $\eta \to \bar{\eta}$, $\bar{\eta} \to \eta$. Semiclassical states ψ of
the given particle (i.e., sections of η) thus correspond bijectively to similar

states of $\bar{\eta}$. However, assignments that are antilinear in ψ as a section of η become linear when $\bar{\eta}$ is considered instead. Since multilinear maps are of fundamental importance for describing interactions as well as composite particles (see §1.10, §5.4), mappings which are antilinear with respect to certain variables can be similarly interpreted as interactions, or composite systems, involving the corresponding antiparticles.

References: Sudbery 1986; Ryder 1986.

1.7 C-parity*

The complex vector bundles η that represent matter fields (to be described in §§1.16, 1.17) are all naturally obtained as complexifications of specific real bundles ζ (their *real forms*), so that $\eta = \zeta \underset{\mathbb{R}}{\oplus} i\zeta$. The states ψ of the particle living in η (§1.2) then can be *complex-linearly* identified with the states $\bar{\psi}$ of its antiparticle in $\bar{\eta}$, where $(\ )^- : \eta \to \bar{\eta}$ is the *complex conjugation isomorphism* equal to Id on $\zeta \oplus 0$ and $-$Id on $0 \oplus i\zeta$. The relation $\psi \leftrightarrow \bar{\psi}$ may be thought of as assigning to any physical state of the particle *the same state* of its antiparticle.

There exist particles in nature whose all properties, i.e., invariants and behavior modes, coincide with those of their antiparticles. This obviously means that the particle in question is its own antiparticle, which one also expresses by calling the particle *strictly neutral*. Letting *generalized charges* be those numerical invariants of particles that change sign under antiparticle formation, with the electric charge as an example (§2.13, §4.14), we thus conclude that they all vanish for a strictly neutral particle, which in particular has to be electrically neutral. Since ψ and $\bar{\psi}$ as above then are the same state of the same particle, we have $\bar{\psi} = z\psi$ with $z \in \mathbb{C}$ (cf. (b) of §1.2) and, as we want the ψ to form a vector space, $z = \pm 1$ must be the same for all ψ in question.

Thus, states of strictly neutral particles are represented by sections of *real*, rather than complex, vector bundles over the given spacetime $(\mathcal{M}, g)$. The number $z = \pm 1$ then is called the *C-parity* (or *charge conjugation parity*) of the given strictly neutral particle species, and indicates whether it is more natural to describe the particle using the real or imaginary form (ζ or $i\zeta$) of the complexified bundle η.

Examples of strictly neutral particles are the photon γ, the neutral weak boson Z^0, the gluons, and many hadrons, e.g., the neutral pion π^0 (§§6.1, 6.2, 6.5).

Reference: Ryder 1986.

1.8 Generalizations of Particles

Let the vector bundles $\eta_1, \ldots, \eta_k$ (with some additional structures) over the given spacetime $(\mathcal{M}, g)$ represent k distinct particle species. The direct sum $\eta_1 \oplus \ldots \oplus \eta_k$ then corresponds to the simplest common *generalization* of these k particles, which may be thought of as something similar, but not identical, to a particle species. Semiclassical states of this *generalized particle*, i.e., sections ψ of $\eta_1 \oplus \ldots \oplus \eta_k$, are *mixtures* (superpositions) $\psi = \psi_1 \oplus \ldots \oplus \psi_k$ of uniquely determined states of each individual particle, among which one distinguishes the *pure* states representing unique species (with all but one of $\psi_1, \ldots, \psi_k$ vanishing identically). For more details, see (i) of §1.19.

Conversely, if a given bundle η with some geometric structure over $\mathcal{M}$, considered as a potential description of an elementary particle, turns out to be decomposable into a direct sum $\eta = \eta_1 \oplus \ldots \oplus \eta_k$ in a manner compatible with its geometry and invariant under all "naturally" involved transformations (while the summands η_j admit no further decompositions of this kind), we obviously conclude that η represents not a single particle species, but rather k species, corresponding to the summands η_j, for which η is a common generalization.

1.9 Generalized Spin Symmetry*

Particularly interesting cases of particle generalizations (§1.8) are obtained when the k particle species in question are similar enough for the bundles $\eta_1, \ldots, \eta_k$ representing them to be naturally isomorphic. This usually results in much additional freedom of choosing the decomposition of $\eta = \eta_1 \oplus \ldots \oplus \eta_k$, i.e., selecting the η_j as specific subbundles of η. Summands of η other than the given η_j then may be thought of as "mixtures" of the particle species corresponding to the η_j.

As an example, let us consider the *generalized spin symmetry*, which includes nonrelativistic spin, isospin, as well as color and flavor symmetries for quarks (see §§2.10, 2.11, 6.2, 8.3). Suppose there is a fixed vector bundle η_0 over $\mathcal{M}$, isomorphic to each of $\eta_1, \ldots, \eta_k$, which we regard as representing a "generic" particle species that replaces those corresponding to the η_j (hence it is not a particle in the usual sense as for it only those invariants are defined that have equal values for all the η_j). Also, let $\imath$ be a *generalized spin bundle*, i.e., a real or complex vector bundle of fibre dimension k over $\mathcal{M}$, endowed with some additional geometric structure, which is trivial as a bundle with that structure, without having a single "preferred" system of trivializing sections.

The "generalized particle" corresponding to η_0 and $\imath$ then is represented by the tensor product $\eta = \eta_0 \otimes \imath$. *Every* splitting $\imath = \imath_1 \oplus \ldots \oplus \imath_k$ of $\imath$ into trivial line bundles, compatible with its geometry, leads to a decomposition

$\eta = \eta_1 \oplus \ldots \oplus \eta_k$ with $\eta_j = \eta_0 \otimes \imath_j$ isomorphic to η_0. More special situations arise when

(a) The splittings in questions are given by systems $\phi_1, \ldots, \phi_k$ of trivializing sections for $\imath$, so that $\psi \mapsto \psi \otimes \phi_j$ is an explicitly defined isomorphism $\eta_0 \to \eta_j$,

or

(b) $\imath$ with its geometry is given as a product $\imath = \mathcal{M} \times V$ with a vector space V and we restrict our consideration to *constant* splittings $\imath = \imath_1 \oplus \ldots \oplus \imath_k$, i.e., ones where $\imath_j = \mathcal{M} \times L_j$, $L_j \subset V$ being a line.

References: Ryder 1986; Sudbery 1986.

1.10 Composite Particles

Our notion of an "elementary" particle includes objects that may be truly elementary (photons, electrons, neutrinos, quarks, etc.), as well as systems whose structure is obviously composite: nucleons, which are assemblies of quarks; nuclei, atoms, or even molecules. This unified approach is based both on the fact that ignoring much of the internal structure of composite systems still leads to reasonably good approximations, and on the *lack of absolute certainty* about the elementary character of any given particle. (Note that atoms were considered elementary through most of the 19th century.)

We may thus try to obtain particles by putting together representatives of some given particle species, corresponding as in §1.2 to vector bundles $\eta_1, \ldots, \eta_k$ over the fixed spacetime $(\mathcal{M}, g)$. We do not assume these particle species to be all distinct, and completely ignore the questions whether the particles involved are "willing" to stay together at all, even for a shortest time, and, if so, what kind of force holds them together; these questions will be addressed later in §§5.6, 6.1, 6.2. One can plausibly argue (see (iii) of §1.19) that the states of the resulting composite particle will all be sections of the tensor product $\eta_1 \otimes \ldots \otimes \eta_k$. However, we cannot in general expect $\eta_1 \otimes \ldots \otimes \eta_k$ to be *the* bundle representing that particle, since it often admits natural direct-sum decompositions, and then (cf. §1.8) it should rather be thought of as a description of a common generalization of several particle species. In other words, putting particles of some given kinds together may lead to many different results. Thus, the algebraically natural operation $(\psi_1, \ldots, \psi_k) \mapsto \psi_1 \otimes \ldots \otimes \psi_k$ of combining states ψ_j of the constituent particles, with the purpose of obtaining a state of the composite particle, has to be composed with the projection of $\eta_1 \otimes \ldots \otimes \eta_k$ onto a specific direct sumand η. Consequently, we obtain a nonzero vector bundle morphism

$$\eta_1 \otimes \ldots \otimes \eta_k \to \eta \qquad (1.1)$$

which is surjective, as η admits no further natural decomposition, and defined *naturally* (i.e., functorially) in terms of the geometries of $\eta_1, \ldots, \eta_k$. More precisely, (1.1) is *of first order*, i.e., is obtained from the parts of the geometries of $\eta_1, \ldots, \eta_k$ and $T\mathcal{M}$ that are defined *fibrewise*. (Thus, (1.1) may involve the spacetime metric g, but not, for instance, its curvature.)

A composite particle we were trying to describe is, consequently, characterized by a vector bundle η and a morphism (1.1) with the stated properties. We need not separately require that η be a natural direct summand of $\eta_1 \otimes \ldots \otimes \eta_k$, as this follows, up to a natural isomorphism, from (1.1).

1.11 Bosons and Fermions

Suppose now that a few of the particle species $\eta_1, \ldots, \eta_k$ in (1.1) are *identical*. It is an empirical fact that identical particles, in the same internal state and at the same location, are *indistinguishable*. Since constituents of a composite particle already have the same location, it follows that permuting the arguments involved in (1.1) should not affect the physical state (as one only changes the way some identical particles are labeled), and so it should only result in multiplying the image in η by a number (cf. (b) of §1.2). Algebraic properties of permutation groups along with continuity of (1.1) now imply that (1.1), as a k-linear mapping between the fibres over each $x \in \mathcal{M}$, has to be symmetric or skew-symmetric with respect to each group of arguments representing identical particles. It is also reasonable to expect the distinction between symmetry and skew-symmetry to depend on the particle species in question alone (and not on k or the morphism in (1.1)). In this way, every particle is either a *boson*, or a *fermion*, depending on whether several copies of the particle enter composite-system morphisms (1.1) symmetrically, or skew-symmetrically.

Among hadrons, bosons occur more or less as frequently as fermions. All interaction carriers are bosons, while all leptons are fermions. (See §§1.14, 1.15.)

Reference: Sudbery 1986.

1.12 Classification of Particles

Hadrons which are bosons (or fermions) are referred to as *mesons* (or, *baryons*). Furthermore, baryons are naturally divided into the disjoint classes of *baryons proper* and *antibaryons*, the latter being the antiparticles of the former (see §2.13). The classification of elementary particles, outlined in §1.5, can now be summarized as follows:

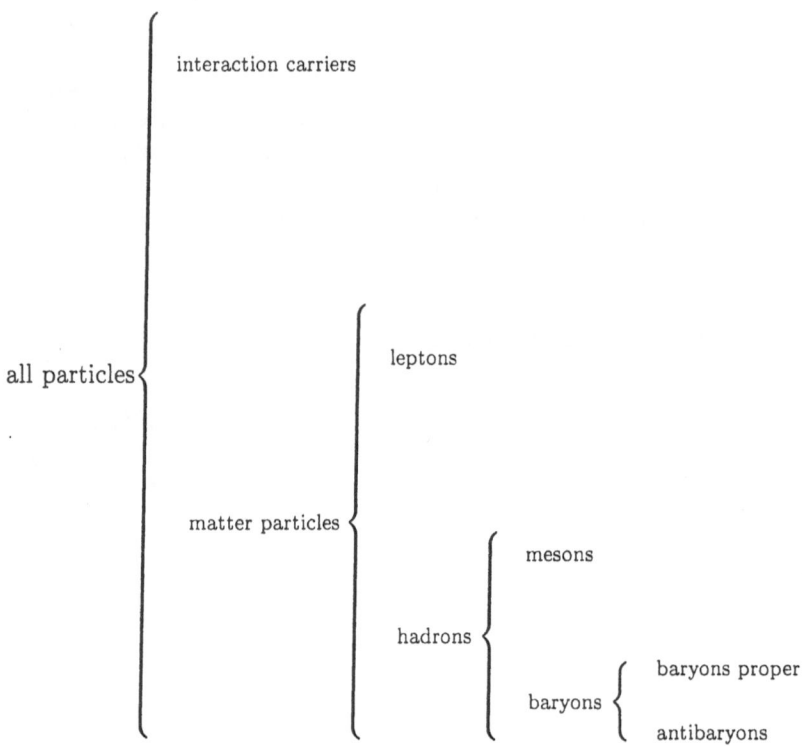

Reference: Ryder 1985.

1.13 Orbital Angular Momentum

Putting k particles together (with $\eta_1, \ldots, \eta_k$ as in §1.10), one can also try to make the resulting system rotate, i.e., endow the particles with some angular momentum about their common center of mass. Discreteness of the spectrum and related properties of the angular-momentum component operators in Euclidean 3-space (cf. §2.7) now make it plausible to regard the angular momentum in question as coming in quantities of the form $\ell\hbar$, with $\ell = 0, 1, 2, \ldots$ and $\hbar$ equal to the Planck constant divided by 2π. Moreover, this quantity of "ℓ units" of angular momentum turns out to behave, formally, as a particle living in the bundle $S_0^\ell T^*\mathcal{M}$ the fibre of which over $x \in \mathcal{M}$ consists of all elements of the ℓth symmetric power

$S^\ell T_x^* \mathcal{M}$ of $T_x^* \mathcal{M}$ whose g-contraction is zero, i.e., of all degree ℓ pseudo-spherical harmonics in $T_x \mathcal{M}$. Again, the composite particle (with ℓ units of angular momentum) is characterized by a bundle morphism

$$S_0^\ell T^* \mathcal{M} \otimes \eta_1 \otimes \ldots \otimes \eta_k \to \eta \qquad (1.2)$$

with similar properties as for (1.1).

1.14 Spin and Parity

In macroscopic physics the angular momentum is a universally conserved quantity. Cases where its conservation seems violated in the microworld can be "explained away" by assuming that each elementary particle species has some capacity of carrying angular momentum in an "internal" manner, independent of the reference point, as if it were spinning about its own axis. This capacity is measured by the *spin* of the particle, which is a non-negative integer or half-integer. The possible measured values of the angular-momentum component in any given direction in 3-space are $0, \hbar, 2\hbar, 3\hbar, \ldots$, while the corresponding spin component for a particle of spin s has the finite spectrum (set of values)

$$-s\hbar, (-s+1)\hbar, \ldots, (s-1)\hbar, s\hbar.$$

See §2.7. The spin of charged particles is also manifested through magnetic effects, which is how the spin was first discovered (for electrons).

Nonrelativistic *quantization* (§2.5) of a system of n classical pointlike particles, moving in the affine Euclidean 3-space $\mathbb{M}$, leads to the Hilbert space $\mathcal{X}$ of all L^2 sections of a vector bundle ζ over the n-fold Cartesian product $\mathbb{M}^n = \mathbb{M} \times \ldots \times \mathbb{M}$. Observables (physical quantities) then are self-adjoint, densely defined operators in $\mathcal{X}$, while the set of possible values the physical quantity may assume is just the spectrum of the corresponding operator. In all relevant cases, sections of ζ may be naturally identified with functions on $\mathbb{M}^n$ valued in a suitable vector space and, in particular, they may be composed with transformations of $\mathbb{M}^n$. In this way, for any fixed point $\mathbf{x}_0 \in \mathbb{M}$ one defines the observable of *orbital parity* about $\mathbf{x}_0$, which is the involutive, isometric, self-adjoint operator in $\mathcal{X}$ acting on vector-valued functions by composing them with the transformation that acts on each $\mathbb{M}$ factor of $\mathbb{M}^n$ as the reflection about $\mathbf{x}_0$. Orbital parity about $\mathbf{x}_0$ may be thought of as a remote analogue of the angular momentum about $\mathbf{x}_0$ (§2.8). In particular, both are usually discussed for $\mathbf{x}_0$ which is the systems's *center of mass*. (On the other hand, the spectrum of parity consists of $1, -1$ only.)

Parity fails to be conserved (i.e., remain unchanged) in weak interactions. Its apparent violation in strong and electromagnetic interactions can, however, be averted by assuming that the participating particles have their "intrinsic" *parities* (analogues of "intrinsic angular momentum", i.e., spin), equal to 1 or -1. The *total parity* of the system in each eigenstate of the operator of parity about the center of mass then is defined to be the product of the orbital and all intrinsic parities involved, and it is this quantity that strong and electromagnetic interactions leave unchanged, even if some particles are created or destroyed in the process. See also §2.8.

It is an empirical fact (corroborated by a theoretical argument known as the *spin-statistics theorem*) that *all bosons have integral spins, while the spins of all fermions are (non-integral) half-integers*. Under antiparticle formation (§1.6), the spin remains unchanged, whereas the parity is preserved for bosons and reversed for fermions (§2.2 and end of §2.4). Boson parity can always be determined, while for fermions, parity is largely conventional as experiments can only determine *relative parities* of some pairs of fermions. Thus, to define parities of baryons, one has to *assume* that p, n and the Λ hyperon have parity $+1$.

There are twelve known leptons: The electron e^-, muon μ^-, tauon τ^-, with positive masses, the massless neutrinos ν_e, ν_μ, ν_τ (electronic, muonic, tauonic), and the corresponding antiparticles $e^+, \mu^+, \tau^+, \bar{\nu}_e, \bar{\nu}_\mu, \bar{\nu}_\tau$. They all have spin $\frac{1}{2}$, while their parities are undefined. Using formal models one usually assumes that the parity is $+1$ for the massive "leptons proper" e^-, μ^-, τ^- and -1 for the massive "antileptons" e^+, μ^+, τ^+, although this can be neither proved nor disproved by experiment. As for neutrinos or antineutrinos, the models describing them contain no data on parity at all, since each neutrino species *determines an orientation of space* and hence cannot undergo space reflections. (See also §1.16.)

References: Ryder 1986; Sudbery 1986.

1.15 The Lightest Hadrons*

The lightest baryons and mesons along with their masses, spins and parities are listed in the following tables, where each particle species is represented by a dash ———— . In parentheses, next to the symbol of each multiplet (cluster) of hadron species, stand the electric charges occurring in the multiplet, in units of proton charge, followed by the average mass of a multiplet member (in MeV/c^2, rounded off to an integer).

As an example, N(0, +; 939) in Table 1.15.2 denotes the *nucleon doublet*, formed by the neutron n $= n^0$ and the proton p $= p^+$ with the approximate masses, in MeV/c^2, given by $m_n = 939.5731 \pm 0.0027$ and $m_p = 938.2796 \pm 0.0027$.

For more details, see §2.9 and §6.2.

Table 1.15.1. The lightest mesons

mass MeV/c^2	spin 0 parity −	spin 0 parity +	spin 1 parity −	spin 1 parity +
100 −				
	$\pi(-,0,+;138)$			
200 −				
300 −				
400 −				
	K(0,+; 496) $\overline{\text{K}}$(0,−; 496)			
500 −				
	$\eta(0; 549)$			
600 −				
700 −				
			$\rho(-,0,+;770)$	
800 −			$\omega(0; 783)$	
900 −			K*(0,+; 892) $\overline{\text{K}^*}$(0,−; 892)	
		S(0; 975)		
1000 −	$\eta'(0; 958)$	$\delta(-,0,+; 980)$		
			$\phi(0; 1020)$	
1100 −				
1200 −				H(0; 1190)

Table 1.15.2. The lightest baryons

mass MeV/c^2	spin 1/2 parity +	spin 1/2 parity −	spin 3/2 parity +	spin 3/2 parity −
900—				
	N(0,+; 939)	$\overline{N}$(0,−; 939)		
1000—				
1100—				
	Λ(0; 1116)	$\overline{\Lambda}$(0; 1116)		
1200—	Σ(−,0,+; 1193)	$\overline{\Sigma}$(−,0,+; 1193)		
			Δ(−,0,+,++; 1232)	$\overline{\Delta}$(+,0,−,−−; 1232)
1300—	Ξ(−,0; 1318)	$\overline{\Xi}$(+,0; 1318)		
1400—	$\overline{\Lambda}$(0; 1405)	Λ(0; 1405)	Σ(−,0,+; 1385)	$\overline{\Sigma}$(−,0,+; 1385)

Reference: Ryder 1986.

1.16 Models of Free Matter Particles with Spins 0 and 1/2

We now proceed to describe the vector bundles η that serve as models, in the sense of §1.2, of matter particles with spins $s \leq \frac{1}{2}$. All particles discussed here are assumed to be *free*, i.e., their interactions are ignored. The choice of the bundles and field equations is justified by the general arguments at the end of §1.18 and the more detailed ones given in §3.9.

A particle of mass $m \geq 0$, the states of which are represented by sections ψ of a vector bundle η with a fixed connection $\overset{\circ}{\nabla}$ over a given spacetime $(\mathcal{M}, g)$, is expected, for reasons given at the end of this section, to satisfy the *Klein-Gordon equation*

$$\left(\square - \frac{m^2 c^2}{\hbar^2} \right) \psi = 0, \tag{1.3}$$

where $c, \hbar$ are, respectively, the speed of light in the vacuum and Planck's constant h divided by 2π, while $\square$ is the *d'Alembertian (wave operator)*, determined by $\overset{\circ}{\nabla}$ and the metric g (as well as its Levi-Civita connection), and assigning to ψ the g-contraction $\square\,\psi$ of its second covariant derivative $\overset{\circ}{\nabla}^2\psi = \overset{\circ}{\nabla}(\overset{\circ}{\nabla}\psi)$. More precisely, $\square\,\psi$ is obtained by applying to the section $\overset{\circ}{\nabla}(\overset{\circ}{\nabla}\psi)$ of $T^*\mathcal{M} \otimes T^*\mathcal{M} \otimes \eta$ the composite bundle morphism

$$T^*\mathcal{M} \otimes T^*\mathcal{M} \otimes \eta \overset{g \otimes \mathrm{Id}}{\underset{\cong}{\longleftarrow}} T\mathcal{M} \otimes T^*\mathcal{M} \otimes \eta = \mathrm{End}(T\mathcal{M}) \otimes \eta \overset{\mathrm{Trace} \otimes \mathrm{Id}}{\longrightarrow} \eta.$$

For particles of spin 0, the Klein-Gordon equation forms a complete description of the "evolving state" ψ. In other words, particles of spin 0 and parity $+1$ live in the product line bundle $\eta = \mathcal{M} \times \mathbb{C}$ (or $\mathcal{M} \times \mathbb{R}$, if the particle is strictly neutral) and are governed by the Klein-Gordon equation (1.3) (where ψ now is a function). On the other hand, particles of spin 0 and parity -1 are represented by the line bundle $\Lambda^4 T^*\mathcal{M}$ of highest degree exterior forms, i.e., pseudoscalars (in the strictly neutral case), or by its complexification $\Lambda^4 T^*\mathcal{M} \otimes \mathbb{C}$ (for particles that are not strictly neutral) and, again, "evolve" according to the Klein-Gordon equation.

Recall (cf. §1.6) that a particle species is *strictly neutral* if it coincides with its own antiparticle. Such particles are naturally described in terms of *real* rather than complex bundles.

Any Weyl or Majorana spinor bundle σ over a spacetime $(\mathcal{M}, g)$ (oriented in the Weyl case) admits a unique connection $\overset{\circ}{\nabla}$, known as the *spinor connection*, which along with the Levi-Civita connection of $(\mathcal{M}, g)$ makes all ingredients of the geometry of σ listed in §1.3 (i.e., Ω, γ, or $\pm I, \omega, \gamma$) parallel. The *Dirac operator $\mathcal{D}$*, acting on sections ψ of σ, then is defined by

$$\mathcal{D}\psi = \mathbf{c}(\overset{\circ}{\nabla}\psi), \tag{1.4}$$

c being the real vector-bundle morphism $T^*\mathcal{M} \otimes \sigma \rightarrow \sigma$ of *Clifford multiplication*, characterized by

$$\mathbf{c}(p \otimes \phi) = \gamma(p)\phi. \tag{1.5}$$

Particles of spin $\frac{1}{2}$ which happen to distinguish an orientation of space, and hence also of the spacetime, namely, the neutrinos ν_e, ν_μ, ν_τ (and the antineutrinos $\overline{\nu}_e, \overline{\nu}_\mu, \overline{\nu}_\tau$), live in a fixed Weyl spinor bundle σ^W over the oriented spacetime $(\mathcal{M}, g)$ (or, respectively, in its conjugate $\overline{\sigma^W}$, which is a Weyl spinor bundle for $(\mathcal{M}, g)$ with the *opposite orientation*), and are governed by *Weyl's equation*

$$\mathcal{D}\psi = 0. \tag{1.6}$$

Similarly (for reasons stated in §3.9), all remaining particles of spin $\frac{1}{2}$ are represented by the complexification $\sigma = \sigma^M \otimes \mathbb{C}$, or its conjugate $\overline{\sigma}$, of a fixed Majorana spinor bundle σ^M over $(\mathcal{M}, g)$ (with no preferred orientation), and evolve according to *Dirac's equation*

$$\left(\mathcal{D} + \frac{mc}{\hbar}\right)\psi = 0, \tag{1.7}$$

m being the mass of the particle. The bundle $\sigma = \sigma^M \otimes \mathbb{C}$ is called the *Dirac spinor bundle* corresponding to σ^M, and its sections ψ are referred to as *Dirac spinor fields* (or *Dirac spinors*).

We have the *Lichnerowicz formula* (see [Besse 1987])

$$\mathcal{D}^2 = \square - \frac{1}{4}\mathrm{Scal}, \tag{1.8}$$

with the minus sign due to the fact that, according to standard notational conventions, our spinors correspond to the metric $-g$. Therefore, Dirac's equation (1.7) does *not* imply the Klein-Gordon equation (1.3) with the same m unless the scalar curvature function Scal of $(\mathcal{M}, g)$ vanishes identically. (Also, if Scal is constant and sufficiently large compared to m, Dirac's equation with mass m implies the Klein-Gordon equation with a different mass.) This discrepancy can be either ignored, or swept under the rug by assuming that Scal = 0 identically, which is a relatively mild restriction on the geometry of $(\mathcal{M}, g)$, satisfied in particular by the Minkowski spacetime. Another possibility is to replace the Klein-Gordon equation (1.3) by the equality

$$\left(\square - \frac{m^2 c^2}{\hbar^2} - \frac{1}{4}\mathrm{Scal}\right)\psi = 0, \tag{1.9}$$

which *does* follow from Dirac's equation (1.7) in view of (1.8). Equation (1.9) is an acceptable substitute, since it has the correct "classical limit" (1.10), as explained below.

The reason why the Klein-Gordon equation is expected to hold is that it is the "simplest" quantized version (§4.8) of the relativistic energy-momentum relation (cf. §4.3)

$$\langle p, p \rangle + m^2 c^2 = 0, \tag{1.10}$$

p being the 4-momentum of a point-like object of mass m, encompassing its total relativistic energy and relativistic 3-momentum. However, quantization of classical relations is a multiple-choice procedure, due to commutativity of classical observables and its lack on the quantum level. In spinor bundles, a different order of factors in the monomials forming the coordinate expression of (1.10) will actually result, instead of the Klein-Gordon equation (1.3), in Eq. (1.9).

References: Schweber 1962; Ramond 1981; Leite Lopes 1981; Hayward 1976; Bogolyubov and Shirkov 1983.

1.17 Higher Spins

The choice of bundles and field equations for free matter particles of spins $s \geq 1$, described below, will be justified in §3.9. See also end of §1.18.

A particle of spin $s = k \in \mathbb{Z}$ and parity $(-1)^k$ lives in the bundle $\eta = S_0^k T^* \mathcal{M}$ (see §1.12), if it is strictly neutral, and in its complexification $S_0^k T^* \mathcal{M} \otimes \mathbb{C}$, if it is not. Furthermore, particles of spin $k \in \mathbb{Z}$ and parity $(-1)^{k+1}$ correspond to the real vector bundle $S_0^k T^* \mathcal{M} \otimes \Lambda^4 T^* \mathcal{M}$, or its complexification $S_0^k T^* \mathcal{M} \otimes \Lambda^4 T^* \mathcal{M} \otimes \mathbb{C}$, depending on whether they are, or are not, strictly neutral. In all these cases, the field equations consist of the Klein-Gordon equation (1.3) (with the mass m of the particle) along with the *divergence condition*

$$\operatorname{div} \psi = 0, \tag{1.11}$$

where $\operatorname{div} \psi$ is the g-contraction of $\overset{\circ}{\nabla} \psi$ involving the differentiation argument and one of the $T^* \mathcal{M}$ arguments, while $\overset{\circ}{\nabla}$ stands for the Levi-Civita connection. Therefore, the divergence condition is vacuous if $k = 0$.

Let σ be a fixed Dirac spinor bundle over the spacetime $(\mathcal{M}, g)$. Particles with spins $s = k + \frac{1}{2}$ with $k \in \mathbb{Z}$, $k \geq 1$, live in the subbundle η of $S_0^k T^* \mathcal{M} \otimes \sigma$ (or of its conjugate) given by $\eta = \operatorname{Ker} F$, where F is the Clifford multiplication (1.5) applied to σ and one of the $T^* \mathcal{M}$ arguments. With the inclusion $S_0^k T^* \mathcal{M} \subset \bigotimes^k T^* \mathcal{M}$, η is the intersection of $S_0^k T^* \mathcal{M} \otimes \sigma$ with the kernel of the morphism $\bigotimes^k T^* \mathcal{M} \otimes \sigma \to \bigotimes^{k-1} T^* \mathcal{M} \otimes \sigma$ characterized by $p_1 \otimes \ldots \otimes p_k \otimes \phi \mapsto p_1 \otimes \ldots \otimes p_{k-1} \otimes \gamma(p_k) \phi$. The field equations consist of the Dirac equation (1.7) and the divergence condition (1.11), applied to the σ and $S_0^k T^* \mathcal{M}$ factors, respectively. (More precisely, one has to replace $\mathcal{D}$ in (1.7) by the corresponding *twisted Dirac operator*. See Remark 5.4.2.)

When the spin s equals $\frac{3}{2}$ or 1, the field equations just described are referred to as the *Rarita-Schwinger* and *Proca* equations. Cf. also (10.10).

References: Schweber 1962; Ramond 1981; Leite Lopes 1981; Hayward 1976; Bogolyubov and Shirkov 1983.

1.18 Natural Bundles

Beside relativistic models of particles, discussed so far, *nonrelativistic ones* are also in use. Due to their simplicity, they may be a valuable source of intuition and, on the technical level, lead to relatively easy, simplified calculations. In the nonrelativistic description, the states of a particle of the given kind are the sections ϕ of a vector bundle ζ over a 3-dimensional Riemannian manifold $(\mathbb{M}, g)$ (cf. (a), (b) of §1.2). The latter is obviously to be thought of as a model of the physical space and, in most specific cases, is assumed to be a Euclidean affine 3-space. For more details, see §2.4.

Whether relativistic or not, such a description of the given particle species is expected to lead to the states (i.e., sections of the corresponding bundle $\eta \to \mathcal{M}$, as in §1.2, or $\zeta \to \mathbb{M}$) over the space(time), which resemble *genuine physical objects* in the sense that one can move them around in space, or let them just sit and age, i.e., "travel in time". In other words, in the case where $(\mathcal{M}, g)$ or $(\mathbb{M}, g)$ is affine, we should be given, as a part of the geometry of η or ζ, an operation of lifting isometries of the base manifold to transformations of the total space of the bundle. One expresses this by saying that the bundle should be *natural*, at least with respect to isometries. Moreover, we may allow the lifting to be defined only up to multiplication by nonzero scalars (cf. (b) of §1.2) and, in the case of the Minkowski spacetime, restrict our consideration to isometries that *preserve the time orientation*, since we have no reason to reverse the time flow in physical processes. However, isometries that change the space orientation are acceptable, unless we are dealing with particle species (the neutrinos) that lead to a preferred orientation of space.

The simplest version of "naturality" is represented by *first-order* natural bundles (from now on just called 'natural' for brevity), in which lifting proceeds through lifting of linear isometries between any two tangent spaces of the base manifold to an isomorphism of the corresponding fibres. An additional advantage gained in using this notion comes from the fact that it makes sense for an arbitrary space $(\mathbb{M}, g)$ or spacetime $(\mathcal{M}, g)$, including those that admit no nontrivial isometries. Using the generic notation $\mathcal{U}$ for $\mathbb{M}$ or $\mathcal{M}$ with $n = \dim \mathcal{U} \in \{3, 4\}$ and η for ζ or η, we thus define a *natural bundle* over $(\mathcal{U}, g)$ to be a complex vector bundle $\eta \to \mathcal{U}$ along with a differentiable, possibly multiple-valued functor assigning an isomorphism $\eta_x \to \eta_y$ of the fibres to each linear isometry $T_x\mathcal{U} \to T_y\mathcal{U}$ (preserving the time orientation if $n = 4$) for $x, y \in \mathcal{U}$. Restricting this *lifting functor* to self-maps of $T_x\mathcal{U}$ we obtain a (possibly multiple-valued) representation $G_x \rightarrow \text{Aut}\,\eta_x$ of the corresponding group ($G_x = \text{O}(T_x\mathbb{M})$ if $n = 3$, and $G_x = \text{O}^\uparrow(T_x\mathcal{M})$, defined as in §3.3, if $n = 4$).

Every natural bundle η over $(\mathcal{U}, g)$ is a direct sum of subbundles invariant under all lifts $\eta_x \rightarrow \eta_y$ which are *irreducible*, i.e., admit no further invariant decompositions. In fact, an irreducible decomposition of a fixed fibre η_x under G_x can be uniquely propagated to all η_y using the lifts $\eta_x \rightarrow \eta_y$. On the other hand, since all representations of the identity component of G_x are unimodular, the highest exterior power $\Lambda^k\eta^*$ (where k is the fibre dimension of η) admits a nowhere-zero section Θ, which may be double-valued (i.e., defined at each point up to a sign), and is invariant, up to a sign, under all lifts. Clearly, Θ is unique up to a constant factor and, in the irreducible case, η is admits a unique connection $\overset{\circ}{\nabla}$ that, along with the Levi-Civita connection of g, makes Θ and the family of representations $G_x \rightarrow \text{Aut}\,\eta_x$, parallel.

The representation $G_x \rightarrow \text{Aut}\,\eta_x$ at any fixed $x \in \mathcal{U}$, if irreducible, determines the whole natural bundle up to a lifting-preserving bundle isomorphism. The isomorphism can namely be chosen first between the fibres over x, and then propagated uniquely under the requirement that Θ be preserved.

As remarked in §1.8, it is the *irreducible* natural bundles that may represent individual species of matter particles. These bundles, over any given $(\mathcal{U}, g)$, are now classified by (single- or multiple-valued) irreducible, finite dimensional complex representations of the group G_x, isomorphic to O(3) if $n = 3$, or to $\text{O}^\uparrow(3,1)$ if $n = 4$. For a classification of these representations, see §2.2 and §3.7.

Consequently, the (first-order) natural bundles are nothing else than the bundles associated with the orthonormal-frame bundle of $(\mathcal{U}, g)$, or with a spin structure over $(\mathcal{U}, g)$, via a linear representation of the structure group.

One of the reasons why the bundles described in §1.16 and §1.17 are used to represent free matter particles is that they are all irreducible and natural (provided that the real bundles are either excluded from consideration, or taken care of by a more general definition of naturality). See also §3.9.

References: Palais and Terng 1977; DeWitt 1965.

1.19 The Physical Meaning of Vector Bundle Operations*

We now proceed to justify the physical interpretations of $^-$, $\oplus$ and $\otimes$ given in §§1.6, 1.8 and 1.10. Since these operations are applied fibrewise, one may replace vector bundles over a spacetime $(\mathcal{M}, g)$ by bundles over a Riemannian 3-manifold $(\mathbb{M}, g)$, i.e., nonrelativistic particle models in the sense of §1.18. We can obviously think of $\mathbb{M}$ as a spacelike submanifold of $\mathcal{M}$, consisting of all events that some observer perceives as taking place at a fixed time t_0 (cf. §4.2).

(i) Let k distinct particle species live in vector bundles $\zeta_1, \ldots, \zeta_k$ over $\mathbb{M}$. The space of sections of $\zeta = \zeta_1 \oplus \ldots \oplus \zeta_k$, interpreted as states of an object living in ζ (§1.18), then contains all states of the given k species, and is spanned by them. The object in question is, therefore, the "smallest" one including these species as its special cases, and so it may be regarded as their common generalization.

(ii) A *combined system* formed by k particles of (not necessarily distinct) species living in vector bundles $\zeta_1, \ldots, \zeta_k$ over $\mathbb{M}$ is obtained simply by *thinking* of these k particles as an assembly, without even requiring that they stay close to one another. The states of such a combined system then are sections of the bundle $\zeta = \zeta_1 \times \ldots \times \zeta_k$ over the kth Cartesian power $\mathbb{M}^k$ of $\mathbb{M}$, with the fibres $\zeta_{(\mathbf{x}_1, \ldots, \mathbf{x}_k)} = (\zeta_1)_{\mathbf{x}_1} \otimes \ldots \otimes (\zeta_k)_{\mathbf{x}_k}$, where $\mathbf{x}_j \in \mathbb{M}$. This can be justified in the following three steps.

a) $\mathbb{M}^k$ is obviously the *configuration space* (manifold of all positions, §2.5) of the combined system.

b) In the simplest case where all particles involved have no internal structure, each ζ_j is a product line bundle, e.g, $\mathbb{M} \times \mathbb{C}$ (see end of §1.2), and the states of each particle are just scalar functions on the configuration space. By a), the states of the combined system then are functions $\mathbb{M}^k \to \mathbb{C}$. Thus, the space $\mathcal{X}$ of states of the combined system is given by

$$\mathcal{X} = \mathcal{X}_1 \otimes \ldots \otimes \mathcal{X}_k, \tag{1.12}$$

$\mathcal{X}_j$ being the space of states of the jth particle. (This discussion can be made rigorous, e.g., by using Hilbert spaces of L^2 functions, as in §2.5.)

c) Our assertion now amounts to generalizing formula (1.12) from the case of structureless particles to arbitrary ones.

(iii) With $\zeta_1, \ldots, \zeta_k$ as in (ii), a composite object (particle) formed by the k particles involved is obtained from the corresponding combined system by requiring all the constituents to be at the same location. This amounts to restricting $\zeta_1 \times \ldots \times \zeta_k$ to the diagonal of $\mathbb{M}^k$, naturally diffeomorphic to $\mathbb{M}$. The states of the composite particle are therefore sections of the bundle $\zeta_1 \otimes \ldots \otimes \zeta_k$, which is the result of this restriction.

(iv) In view of (i), (ii), (iii), it is natural to expect all functorial constructions involving vector bundles to have physical interpretations. Thus, one can *define* antiparticle formation (§1.6) to be the physical counterpart of the complex conjugation functor $\zeta \mapsto \bar{\zeta}$.

(v) The dual-bundle functor $\zeta \mapsto \zeta^*$ is, by itself, of little interest as an operation on particles since, in all relevant cases, ζ^* is naturally isomorphic to $\bar{\zeta}$ or ζ. See Remark 6.3.2.

2 Invariants of Particles*

This chapter provides a description of quantum numbers and other numerical invariants associated with elementary particles. Further invariants are discussed in Chapter 8.

2.1 Representations of SU(2) and SO(3)

Let $(S, \Omega, \langle,\rangle)$ be a *Pauli triple*, consisting of a 2-dimensional complex vector space S with a positive definite Hermitian inner product $\langle,\rangle$ and a nonzero 2-form $\Omega \in \Lambda^2 S^*$ which is compatible with $\langle,\rangle$ in the sense that $\Omega(\psi_1, \psi_2) = 1$ for some orthonormal basis ψ_1, ψ_2 of $(S, \langle,\rangle)$. The 3-dimensional real vector space T of all traceless Hermitian endomorphisms of S then carries a natural Euclidean inner product, also denoted $\langle,\rangle$ and given by $\langle B_1, B_2 \rangle = \frac{1}{2}\text{Trace}(B_1 B_2)$ for $B_1, B_2 \in T$. (Actually, setting $B_1 = B_2$, one easily sees that $B_1 B_2 + B_2 B_1 = 2\langle B_1, B_2 \rangle \cdot \text{Id}_S$ for $B_1, B_2 \in T$, which establishes an isomorphism between $\text{End}\, S$ and the Clifford algebra of $(T, -\langle,\rangle)$.) The group $\text{SU}(S) \cong \text{SU}(2)$ of all automorphisms of S preserving Ω and $\langle,\rangle$ has a natural $\mathbb{Z}_2$ extension $\overline{\text{SU}}(S) \subset \text{Aut}_{\mathbb{R}} S$ such that $\overline{\text{SU}}(S) \smallsetminus \text{SU}(S)$ consists of all antiautomorphisms of S sending $\Omega, \langle,\rangle$ onto $\overline{\Omega}, \overline{\langle,\rangle}$.

The antiautomorphism Π of S characterized by

$$\Omega(\psi, \phi) = \langle \psi, \Pi\phi \rangle \qquad\qquad (2.1)$$

for all $\psi, \phi \in S$ satisfies $\Pi\psi_1 = -\psi_2$, $\Pi\psi_2 = \psi_1$ whenever ψ_1, ψ_2 are orthonormal and $\Omega(\psi_1, \psi_2) = 1$. Therefore, $\Pi^2 = -\text{Id}_S$ and so, by skew-symmetry of Ω, $\Pi \in \overline{\text{SU}}(S)$. Clearly, Π commutes with all of $\overline{\text{SU}}(S)$, as it is naturally determined by Ω and $\langle,\rangle$, as well as $\overline{\Omega}$ and $\overline{\langle,\rangle}$. It is also easy to verify, reducing the problem to $\text{SU}(S)$ with the aid of Π, that $\overline{\text{SU}}(S)$ has the center $\{\text{Id}, \Pi, -\text{Id}, -\Pi\}$, isomorphic to $\mathbb{Z}_4$. Moreover, Π anticommmutes with each $B \in T$, which is immediate from the relation $\Omega(B\psi, \phi) + \Omega(\psi, B\psi) = 0$, equivalent to $\text{Trace}\, B = 0$. (In general, we have $\Omega(B\psi, \phi) + \Omega(\psi, B\phi) = (\text{Trace}\, B)\Omega(\psi, \phi)$ for $B \in \text{End}\, S$, ψ, $\phi \in S$.)

The action of $\overline{\text{SU}}(S)$ on T by conjugation gives rise to a homomorphism $\overline{\text{SU}}(S) \rightarrow \text{O}(T)$ into the group of linear isometries of $(T, \langle,\rangle)$, sending $\pm\Pi$ onto $-\text{Id}_T$ (see above) and having the kernel $\{\text{Id}_S, -\text{Id}_S\}$. Hence, for dimensional reasons, this homomorphism is a twofold covering. In particular, one has a twofold covering homomorphism $\mathrm{SU}(S) \to \mathrm{SO}(\mathcal{T})$ between the identity components, so that $\mathrm{SO}(3) \cong \mathrm{SU}(2)/\mathbb{Z}_2$ (see also §3.4).

Since $\mathrm{SU}(2)$ is simply connected (in fact, isomorphic to the sphere group S^3 of unit quaternions, acting on the space $\mathbb{H} \cong \mathbb{C}^2$ of quaternions by left quaternion multiplication, cf. §3.4), its representations correspond bijectively to those of its Lie algebra $\mathfrak{su}(2) \cong \mathfrak{so}(3) \cong \mathbb{R}^3$, the corresponding Lie bracket in $\mathbb{R}^3$ being the vector product. Representations of $\mathfrak{su}(2)$ in finite dimensional complex vector spaces W can be classified as follows. Let $X \in \mathfrak{su}(2)$ have unit norm relative to the Euclidean inner product of $\mathbb{R}^3 \cong \mathfrak{su}(2)$, which is just a suitably normalized Killing form of $\mathfrak{su}(2)$. Hence there exist $Y, Z \in \mathfrak{su}(2)$ with $[X, Y] = Z$, $[Y, Z] = X$, $[Z, X] = Y$. Identifying X, Y, Z with the endomorphisms of W associated with them by a given representation and setting $W_\lambda = \mathrm{Ker}\,(X - \lambda \cdot \mathrm{Id}_W)$ for $\lambda \in \mathbb{C}$ and $\Phi^\pm = Y \pm iZ \in \mathrm{End}\, W$, we thus have

$$[X, \Phi^\pm] = \pm i\Phi^\pm, \quad [\Phi^+, \Phi^-] = -2iX, \quad \Phi^\pm(W_\lambda) \subset W_{\lambda \mp i}.$$

For any fixed $x \in W_{\lambda_0} \setminus \{0\}$, where λ_0 is an eigenvalue of X in W with the largest possible imaginary value, and integers $m \geq 0$, let $x_m = (\Phi^+)^m x$. Then, by induction on m,

$$x_m \in W_{\lambda_0 - mi}, \quad X x_m = (\lambda_0 - mi)x_m, \quad \Phi^+ x_m = x_{m+1}, \\ \Phi^- x_m = m(2i\lambda_0 + m - 1)x_{m-1}. \tag{2.2}$$

The vectors x_m are linearly independent as long as they are nonzero, since then they are eigenvectors of X. As $\dim W < \infty$, there exists n with $x_{n-1} \neq 0$, $x_n = 0$. Let $s = \frac{n-1}{2}$. Condition $\Phi^- x_n = 0$ now implies, in view of the above formula for $\Phi^- x_m$, that $\lambda_0 = si$. Thus, the vectors $x_0, \ldots, x_{n-1}$ span an invariant subspace of dimension $n = 2s + 1$ in the given representation space. According to (2.2), the resulting subrepresentation is uniquely determined by s up to an equivalence, and it is irreducible, since any of its nontrivial invariant subspaces has to contain an eigenvector of X, i.e., a multiple of one of the x_m. The number

$$s = -i\lambda_0 \in \{0, \frac{1}{2}, 1, \frac{3}{2}, \ldots\} \tag{2.3}$$

is therefore an invariant of the original representation of $\mathfrak{su}(2)$ in W, called its *spin* and characterized by $2s + 1$ being the highest dimension of its irreducible subrepresentations or, equivalently, by si being the eigenvalue in W of some (any) unit $X \in \mathfrak{su}(2)$, with the largest possible imaginary value. An irreducible representation is uniquely determined, up to an equivalence, by its spin s, and any unit $X \in \mathfrak{su}(2)$ then has the eigenvalues

$$-si, (-s+1)i, \ldots, (s-1)i, si, \tag{2.4}$$

all simple. Decomposing any representation into irreducible summands and constructing in each of them the eigenvectors x_m as above for a fixed unit

$X \in \mathfrak{su}(2)$, we conclude that *a representation of $\mathfrak{su}(2)$ such that some unit $X \in \mathfrak{su}(2)$ has the eigenvalue $z \in \mathbb{C}$, must contain an irreducible subrepresentation of dimension $n \geq 2|z| + 1$ (and, necessarily, $z \in \frac{1}{2}i\mathbb{Z}$).* This implies the following *criterion of irreducibility: A representation of $\mathfrak{su}(2)$ in W under which some unit $X \in \mathfrak{su}(2)$ has an eigenvalue $z \in \mathbb{C}$ with $\dim W \leq 2|z| + 1$ is necessarily irreducible, with spin $s = |z|$, and $\dim W = 2|z| + 1$.*

Irreducible multi-valued representations of SO(3) are either single, or double-valued (if additional "artificial" branches are removed), since all representations of its simply connected twofold covering group SU(2) are essentially single-valued. Obviously, an irreducible representation of SU(2) descends to a (single-valued) representation of SO(3) if and only if it sends $-\mathrm{Id}$ onto Id, and does not descend (i.e., becomes double-valued on the SO(3) level) if and only if it maps $-\mathrm{Id}$ onto $-\mathrm{Id}$. Such a representation or, equivalently, the corresponding representation of the Lie algebra $\mathfrak{su}(2)$, is called *bosonic* in the former case and *fermionic* in the latter. Bosonic (irreducible) representations are characterized by their spins being integers ($s \in \mathbb{Z}$), while fermionic ones are those with $s \in (\frac{1}{2}\mathbb{Z}) \setminus \mathbb{Z}$. In fact, the circle subgroup $\exp tX$ in SO(3), $0 \leq t \leq 2\pi$, with $X \in \mathfrak{su}(2)$ of norm 1, is mapped under a representation of spin s onto

$$\mathrm{diag}\{e^{-sit}, e^{(-s+1)it}, \ldots, e^{(s-1)it}, e^{sit}\} \tag{2.5}$$

for some diagonalizing basis of the representation space (a basis of eigenvectors for X). The resulting map of the circle $S^1 = \mathbb{R}/2\pi\mathbb{Z}$ is obviously single or double valued, according to whether $s \in \mathbb{Z}$ or $s \in (\frac{1}{2}\mathbb{Z}) \setminus \mathbb{Z}$.

References: Zhelobenko 1973; Mackey 1963; Schweber 1962; Sudbery 1986.

2.2 Representations of O(3)

Let $(S, \Omega, \langle,\rangle)$ and $\mathcal{T}$ be as in §2.1. Irreducible representations of $O(\mathcal{T})$ are, again, single or double-valued, i.e., as representations of $\overline{\mathrm{SU}}(S)$ they send $-\mathrm{Id}_S$ onto Id or, respectively, onto $-\mathrm{Id}$, which we also express by calling them *bosonic* or *fermionic*. Due to irreducibility, the representation sends $-\mathrm{Id}_{\mathcal{T}}$ in the bosonic case (or, Π defined by (2.1), in the fermionic case) onto $\epsilon \cdot \mathrm{Id}$ (or $\epsilon i \cdot \mathrm{Id}$), for a number $\epsilon = \pm 1$ called the *parity* of the representation. Since $\overline{\mathrm{SU}}(S)$ is a quotient group of $\mathrm{SU}(S) \times \{\mathrm{Id}_S, \Pi, -\mathrm{Id}_S, -\Pi\}$, one easily sees that *irreducible finite-dimensional complex representations of $\overline{\mathrm{SU}}(S)$ are classified by the spin $s \in \{0, \frac{1}{2}, 1, \frac{3}{2}, \ldots\}$ and parity $\epsilon \in \{1, -1\}$.* All pairs s, ϵ are in fact realized; see the examples in §2.3.

Parity of fermionic representations (with $s \in (\frac{1}{2}\mathbb{Z}) \setminus \mathbb{Z}$) is actually conventional, since i and $-$i are congruent modulo a field automorphism of $\mathbb{C}$. Also, two fermionic representations of the same spin and opposite parities are "quasi-equivalent" in the sense that an automorphism of the group

$\overline{\mathrm{SU}}(S)$ is also involved. This automorphism may, e.g., be chosen to keep $\mathrm{SU}(S)$ fixed and multiply $\overline{\mathrm{SU}}(S) \smallsetminus \mathrm{SU}(S)$ by $-\mathrm{Id}_S$.

Replacing an irreducible representation by its *complex conjugate* does not change the spin, which is a function of the dimension. As for the parity, conjugation leaves it unchanged for bosonic representations, but reverses it for fermionic representations, since $i \cdot \mathrm{Id}$ becomes $-i \cdot \mathrm{Id}$ under conjugation.

References: Sudbery 1986; Schweber 1962.

2.3 Geometric Realizations

For a Pauli triple $(S, \Omega, \langle, \rangle)$ and the corresponding Euclidean 3-space $\mathcal{T}$ (§2.1), the group $\overline{\mathrm{SU}}(S)$ acts on S as well as on $\mathcal{T}$ (via the homomorphism $\overline{\mathrm{SU}}(S) \to \mathrm{O}(\mathcal{T})$) and, further, on their dual spaces, tensor products and powers, invariant subspaces of these, etc., through maps which are complex linear or antilinear, or real linear (for real spaces such as $\mathcal{T}$). Given an integer $k \geq 0$, let $\Sigma_k(\mathcal{T}) = S_0^k \mathcal{T}^* \subset S^k \mathcal{T}^* \subset \bigotimes^k \mathcal{T}^*$ be the space of degree k spherical harmonics on $\mathcal{T}$ (cf. §1.13). Also, let $\Sigma_{k+\frac{1}{2}}(S)$ be the subspace of $\Sigma_k(\mathcal{T}) \underset{\mathbb{R}}{\otimes} S \subset \bigotimes^k \mathcal{T}^* \otimes S$ obtained by intersecting $\Sigma_k(\mathcal{T}) \underset{\mathbb{R}}{\otimes} S$ with the kernel of the composite $\bigotimes^k \mathcal{T}^* \underset{\mathbb{R}}{\otimes} S \to \bigotimes^k \mathcal{T} \underset{\mathbb{R}}{\otimes} S \to \bigotimes^{k-1} \mathcal{T} \underset{\mathbb{R}}{\otimes} S$, where the first map is induced by the isomorphism $\mathcal{T} \cong \mathcal{T}^*$ coming from the Euclidean inner product in $\mathcal{T}$, and the second (equal to 0 if $k = 0$) is characterized by $B_1 \otimes \ldots \otimes B_k \otimes \psi \mapsto B_1 \otimes \ldots \otimes B_{k-1} \otimes [B_k(\psi)]$.

Example 2.3.1. If $k \in \mathbb{Z}$, the obvious complexified representation of $\mathrm{SO}(\mathcal{T})$ (or $\mathrm{SU}(S)$, or the Lie algebra $\mathfrak{su}(S)$) in $\Sigma_k(\mathcal{T}) \otimes \mathbb{C}$ is irreducible, of spin k. This is immediate from the irreduciblity criterion in §2.1, applied to the harmonic function $\mathcal{T} = \mathbb{R} \oplus \mathbb{C} \ni (a, z) \mapsto z^k \in \mathbb{C}$ along with the relation $\dim_{\mathbb{R}} \Sigma_k(\mathcal{T}) = 2k + 1$, easily established by induction. The representation of $\mathrm{O}(\mathcal{T})$ or $\overline{\mathrm{SU}}(S)$ in $\Sigma_k(\mathcal{T}) \otimes \mathbb{C}$ is therefore irreducible, of spin k and parity $(-1)^k$.

Example 2.3.2. The natural representation of $\mathrm{O}(\mathcal{T})$ or $\overline{\mathrm{SU}}(S))$ in the line $\Lambda^3 \mathcal{T}^*$ proceeds through the homomorphism

$$\mathrm{O}(\mathcal{T}) \overset{\det}{\longrightarrow} \mathbb{Z}_2 = \{\mathrm{Id}_{\Lambda^3 \mathcal{T}^*}, -\mathrm{Id}_{\Lambda^3 \mathcal{T}^*}\},$$

and so the complexified representation in $\Lambda^3 \mathcal{T}^* \otimes \mathbb{C}$ is irreducible, with spin 0 and parity -1.

Example 2.3.3. By Examples 2.3.1, 2.3.2, the obvious representation of $\mathrm{O}(\mathcal{T})$ or $\overline{\mathrm{SU}}(S)$ in $\Sigma_k(\mathcal{T}) \otimes \Lambda^3 \mathcal{T}^* \otimes \mathbb{C}$ is irreducible, of spin k and parity $(-1)^{k-1}$.

Example 2.3.4. For an integer $k \geq 0$, the representation of $SU(S)$ in $\Sigma_{k+\frac{1}{2}}(S)$ is irreducible, of spin $k + \frac{1}{2}$, as one easily verifies by induction.

Example 2.3.5. The representation of $\overline{SU}(S)$ in $\Sigma_{k+\frac{1}{2}}(S)$ with $k \in \mathbb{Z}$ is not complex linear, since $\overline{SU}(S) \smallsetminus SU(S)$ acts antilinearly. The complexification of its underlying real representation is *reducible*, since the representation space $\Sigma_{k+\frac{1}{2}}(S) \otimes \mathbb{C}$ splits into $\text{Ker}(\Pi - i \cdot \text{Id}) \oplus \text{Ker}(\Pi + i \cdot \text{Id})$, where Π, defined by (2.1), is identified with its image under the representation. The resulting subrepresentations of $\overline{SU}(S)$ in the spaces $\text{Ker}(\Pi \mp i \cdot \text{Id}) \subset \Sigma_{k+\frac{1}{2}}(S) \otimes \mathbb{C}$ are irreducible, of spin $k + \frac{1}{2}$ and parities ± 1, which easily follows from Example 2.3.4.

Example 2.3.6. Representations of $\overline{SU}(S)$ in *complex* vector spaces which are fermionic (i.e., send $-\text{Id}$ to $-\text{Id}$) can be *modified* by multiplying transformations corresponding to $\overline{SU}(S) \smallsetminus SU(S)$ by i, and leaving those in $SU(S)$ unchanged. As a result one obtains a representation of a new twofold covering group $\widetilde{SU}(S)$ of $O(T)$, not isomorphic to $\overline{SU}(S)$. (Note that the centers of $\widetilde{SU}(S)$ and $\overline{SU}(S)$ are, respectively, isomorphic to $\mathbb{Z}_2 \oplus \mathbb{Z}_2$ and $\mathbb{Z}_4$.) Studying $\widetilde{SU}(S)$ instead of $\overline{SU}(S)$ does not, however, lead to any essentially new insights, since the modification procedure just described establishes a bijective correspondence between irreducible complex representations of both groups.

Example 2.3.7. A particularly simple model of an irreducible representation of $SU(S)$ with any spin s is provided by the $(2s)$th symmetric power of S. See Remark 8.2.3.

2.4 Nonrelativistic Particle Models

In a nonrelativistic description, particles of matter are represented by irreducible natural vector bundles ζ (see §1.18) over a fixed "space" (Riemannian 3-manifold) $(\mathbb{M}, g)$. Such bundles are uniquely characterized by equivalence classes of the corresponding (single- or double-valued) representations of $O(3) \cong O(T_x \mathbb{M})$, $x \in \mathbb{M}$ (see end of §1.18), which in turn are classified by their spin and parity (§2.2). The geometric examples of §2.3 now lead to the following constructions of ζ for each given spin s and parity ϵ, with ζ uniquely determined by s and ϵ (up to a lifting-functor preserving bundle isomorphism and, if $s \notin \mathbb{Z}$, also up to the choice of a spinor bundle over $(\mathbb{M}, g)$).

For $s \in \mathbb{Z}$, particles of spin s and parity $(-1)^s$ (or $(-1)^{s+1}$) live in the bundle $\zeta = S_0^k T^* \mathbb{M} \otimes \mathbb{C}$ (or, $\zeta = S_0^k T^* \mathbb{M} \otimes \Lambda^3 T^* \mathbb{M} \otimes \mathbb{C}$), cf. §§1.16, 1.17, with the exception of strictly neutral particles, for which the $\mathbb{C}$ factor has to be dropped (as they are represented by *real* vector bundles).

Before discussing fermions ($s \notin \mathbb{Z}$), it is convenient to introduce the following definition. By a *Pauli spinor bundle* over an *oriented* Riemannian 3-manifold $(\mathbb{M}, g)$ we mean a system $(\sigma^P, \Omega, \langle, \rangle, \gamma)$, where σ^P is a complex vector bundle of fibre dimension 2 over $\mathbb{M}$ and $\Omega, \langle, \rangle$ and γ are C^∞ sections of $\Lambda^2 \sigma^*, \sigma^* \otimes \bar{\sigma}^*$ and $\mathrm{Hom}(T^*\mathbb{M}, \mathrm{End}\,\sigma)$, respectively, with $\sigma = \sigma^P$, such that for each $\mathbf{x} \in \mathbb{M}$, $(\sigma_\mathbf{x}^P, \Omega_\mathbf{x}, \langle, \rangle_\mathbf{x})$ is a Pauli triple and $\gamma_\mathbf{x}$ is an orientation-preserving linear isometry of $T_\mathbf{x}^*\mathbb{M}$ onto the corresponding vector space $T = T_\mathbf{x}$ defined as in §2.1. Moreover, T is naturally oriented for any Pauli triple $(S, \Omega, \langle, \rangle)$. In fact, bases ψ_1, ψ_2 of S with $\langle \psi_\alpha, \psi_\beta \rangle = \delta_{\alpha\beta}$ and $\Omega(\psi_1, \psi_2) = 1$ form an orbit of the *connected* group $\mathrm{SU}(2)$, and to each such basis we can continuously assign an orthonormal basis of T, given by the *Pauli matrices*

$$\begin{bmatrix} 1 & 0 \\ 0 & -1 \end{bmatrix}, \begin{bmatrix} 0 & 1 \\ 1 & 0 \end{bmatrix}, \begin{bmatrix} 0 & -i \\ i & 0 \end{bmatrix}.$$

Pauli spinor bundles always exist, since orientability of $\mathbb{M}$ implies parallelizability if $\dim \mathbb{M} = 3$, and they are classified by elements of $H^1(\mathbb{M}, \mathbb{Z}_2)$. More precisely, equivalence classes of such bundles over a given $(\mathbb{M}, g)$ form an affine space over the field $\mathbb{Z}_2$, whose associated translation vector space is $H^1(\mathbb{M}, \mathbb{Z}_2)$.

Finally, Pauli spinor bundles are natural, in a more restricted sense than in §1.18, so that only *orientation-preserving* linear isometries between tangent spaces of the base manifold $(\mathbb{M}, g)$ are considered. Their spin (i.e., the spin of the obvious representation of $\mathrm{SU}(\sigma_\mathbf{x}^P)$ in $\sigma_\mathbf{x}^P$, $\mathbf{x} \in \mathbb{M}$) equals $\frac{1}{2}$. Thus, they may be used to represent particle species that determine an orientation of space and have spin $\frac{1}{2}$, namely, the neutrinos. More precisely, one assumes that a fixed Pauli spinor bundle σ^P over a given oriented $(\mathbb{M}, g)$ corresponds to each of the three neutrino species ν_e, ν_μ, ν_τ (§1.14), while its conjugate $\overline{\sigma^P}$, which naturally is a Pauli spinor bundle for $(\mathbb{M}, g)$ with the *opposite* orientation, hosts the antineutrinos $\bar{\nu}_e, \bar{\nu}_\mu, \bar{\nu}_\tau$.

For fermions that are not partial to a space orientation, the vector bundles involved have to be constructed using a suitably modified version of Pauli spinor bundles. To this end, let us note that a Pauli triple $(S, \Omega, \langle, \rangle)$ can be equivalently represented by a system $(S, I, \omega, \langle, \rangle_0)$ with I, ω as in §1.3 and the Euclidean inner product $\langle, \rangle_0 = \mathrm{Re}\langle, \rangle$ in the real 4-space S, intrinsically characterized by the requirements that $I^*\langle, \rangle_0 = \langle, \rangle_0$ and $\langle \psi_\alpha, \psi_\beta \rangle_0 = \delta_{\alpha\beta}$, $\langle \psi_1, I\psi_2 \rangle = 0$, $\omega(\psi_1, \psi_2) = 2$, $\omega(\psi_1, I\psi_2) = 0$ for some $\psi_1, \psi_2 \in S$. Pauli triples $(S, \Omega, \langle, \rangle)$ defined only up to the conjugation $(S, \Omega, \langle, \rangle) \mapsto (\overline{S}, \overline{\Omega}, \overline{\langle, \rangle})$ are, therefore, in a bijective correspondence with the *Pauli quadruples* of the form $(S, \pm I, \omega, \langle, \rangle_0)$. For such objects, the corresponding Euclidean 3-space T (§2.1) is still well-defined, but no more naturally oriented.

Let $(\mathbb{M}, g)$ be a Riemannian 3-manifold which is *not* oriented (and may even by nonorientable). A *Pauli spinor bundle* over $(\mathbb{M}, g)$ consists

of a real vector bundle σ^P of fibre dimension 4 over $\mathbb{M}$ along with C^∞
sections $\pm I, \omega, \langle,\rangle_0$ and γ of appropriate bundles such that at each $\mathbf{x} \in$
$\mathbb{M}$, $(\sigma_{\mathbf{x}}^P, \pm I_{\mathbf{x}}, \omega_{\mathbf{x}}, (\langle,\rangle_0)_{\mathbf{x}})$ is a Pauli quadruple and $\gamma_{\mathbf{x}}$ is a linear isometry
of $T_{\mathbf{x}}^* \mathbb{M}$ onto the corresponding Euclidean 3-space $\mathcal{T} = \mathcal{T}_{\mathbf{x}}$. (Thus, $\pm I$ is
C^∞ as a double-valued section of $\operatorname{End} \sigma^P$, so that it has local C^∞ branches.
Existence of global single-valued C^∞ branches of $\pm I$ is equivalent to orientability of $\mathbb{M}$.) The bundle morphism $T^* \mathbb{M} \otimes \sigma^P \to \sigma^P$ characterized by
$\mathbf{p} \otimes \psi \mapsto \gamma(\mathbf{p})\psi$ then is called the *Clifford multiplication*.

The antiautomorphism Π defined by (2.1) remains unchanged when the
Pauli triple $(S, \Omega, \langle,\rangle)$ is replaced by its conjugate $(\overline{S}, \overline{\Omega}, \overline{\langle,\rangle})$, so that Π is
also well-defined for Pauli quadruples. The corresponding assignment $\mathbb{M} \ni$
$\mathbf{x} \mapsto \Pi = \Pi_{\mathbf{x}} \in \operatorname{End} \sigma_{\mathbf{x}}^P$ then becomes a natural section Π of $\operatorname{End} \sigma^P$ for
any Pauli spinor bundle σ^P over a Riemannian 3-manifold $(\mathbb{M}, g)$, while its
tensor product with Id is, for each fixed integer $k \geq 0$, an endomorphism of
the complex vector bundle $S_0^k T^* \mathbb{M} \otimes \sigma^P \otimes \mathbb{C}$, also denoted Π and satisfying
$\Pi^2 = -\operatorname{Id}$. Moreover, Π leaves invariant the subbundle ζ of $S_0^k T^* \mathbb{M} \otimes \sigma^P \otimes \mathbb{C}$
obtained by requiring that the Clifford product involving σ^P and one of the
$T^* \mathbb{M}$ arguments be zero (cf. §1.17 and §2.3; if $k = 0$, $\zeta = \sigma^P \otimes \mathbb{C}$). Thus,
$\zeta = \zeta^+ \oplus \zeta^-$ with $\zeta^\pm = \operatorname{Ker}(\Pi \mp i \cdot \operatorname{Id})$. According to Example 2.3.5, ζ^+ and
ζ^- are irreducible natural bundles of spin $k + \frac{1}{2}$ over $(\mathbb{M}, g)$, with parities
$+1$ and -1, respectively. These bundles are assumed to represent particles
of spin $k + \frac{1}{2}$ and the appropriate parity (with a fixed Pauli spinor bundle
σ^P, common to all k and all particles involved). Note that $\overline{\zeta^\pm}$ is naturally
isomorphic to $\zeta^\mp$, as the complex conjugation antiautomorphism of ζ sends
$\zeta^\pm$ onto $\zeta^\mp$. On the other hand, the bundle representing bosons of any given
spin and parity (see the beginning of this section) is naturally isomorphic
to its own complex conjugate, as it is obtained by complexifying a real
bundle. This accounts for the fact that the parities of a given particle and
its antiparticle coincide for bosons and differ for fermions (§1.14).

Fermion parity is generally conventional (§§1.14, 2.2). In the case of
leptons (§1.14) it is not defined at all, even as a "relative" quantity, which is
in part related to the fact that weak interactions may fail to conserve parity
(the "total parity" in the sense of §1.14). More precisely, parity makes no
sense for (anti)neutrinos due to their orientation-fixing property, while any
parity values assigned to the massive leptons $(e^-, \mu^-, \tau^-, e^+, \mu^+, \tau^+)$ would
not lead to consequences verifiable by experiment. Thus, it is natural to
assume that e^-, μ^-, τ^- (or, e^+, μ^+, τ^+) live in $\sigma^P \otimes \mathbb{C}$ (or, in $\overline{\sigma^P \otimes \mathbb{C}}$) for a
fixed Pauli spinor bundle σ^P over the given Riemannian 3-manifold $(\mathbb{M}, g)$.
We allow here the *reducible* natural bundle $\sigma^P \otimes \mathbb{C} = \zeta^+ \oplus \zeta^-$ in order that
the massive leptons have spin $\frac{1}{2}$, but no definite parity, in agreement with
experimental data.

References: Lawson and Michelsohn 1989; Derdzinski 1991b.

2.5 Canonical Quantization

Let a manifold $\mathbb{M}$ be the space of all possible positions (the *configuration space*) of a classical mechanical system. Elements of $T^*\mathbb{M}$ (or $T\mathbb{M}$) are called the *states* of the system and serve as its "initial conditions", uniquely evolving with time according to specific equations of motion (usually in the form of Hamilton's or Euler-Lagrange equations). The system's *observables* (physical quantities) then are functions $T^*\mathbb{M} \to \mathbb{R}$ (in the Hamiltonian formalism), or $T\mathbb{M} \to \mathbb{R}$ (in the Lagrangian formalism). Two particularly important classes of observables are provided by functions $f : \mathbb{M} \to \mathbb{R}$ and vector fields $\mathbf{v}$ on $\mathbb{M}$, which one regards as functions on $T^*\mathbb{M}$ constant (linear) on each fibre. They are referred to as *generalized position* (and *momentum*) *components*.

Example 2.5.1. Let $\mathbb{M}$ be a Euclidean affine 3-space, in which a pointlike object of mass m is allowed to move. At any point $\mathbf{x} \in \mathbb{M}$, the object's *momentum* is defined by $\mathbf{p} = m\dot{\mathbf{x}} \in T_{\mathbf{x}}\mathbb{M} \cong \mathbf{V} \cong T_{\mathbf{x}}^* \cong \mathbf{V}^*$, where $\mathbf{V}$ is the translation space of $\mathbb{M}$ (with identifications due to the affine and Euclidean structures of $\mathbb{M}$). Thus, $T^*\mathbb{M} \cong \mathbb{M} \times \mathbf{V}^* \cong \mathbb{M} \times \mathbf{V}$ can be thought of as the space of all "positions-and-momenta" of the system. The geometry of $\mathbb{M}$ naturally distinguishes the following two types of generalized momentum-component observables:

(a) Any fixed vector $\mathbf{v} \in \mathbf{V}$, regarded as a constant vector field on $\mathbb{M}$, leads to the observable $T^*\mathbb{M} \ni (\mathbf{x}, \mathbf{p}) \mapsto \mathbf{p}(\mathbf{v}) = m\langle \dot{\mathbf{x}}, \mathbf{v} \rangle$ (the component of the *linear momentum* in the direction of $\mathbf{v}$).

(b) Let $a \in \mathfrak{so}(\mathbf{V})$ be a skew-adjoint endomorphism of $\mathbf{V}$. For any fixed $\mathbf{x}_0 \in \mathbb{M}$, the flow of the vector field $\mathbf{v}_{a,\mathbf{x}_0}$ on $\mathbb{M}$ with $\mathbf{v}_{a,\mathbf{x}_0}(\mathbf{x}) = a(\mathbf{x} - \mathbf{x}_0)$ consists of rotations about the line L through $\mathbf{x}_0$ parallel to $\mathrm{Ker}\, a \subset \mathbf{V}$, with the angular velocity $|a| = \sqrt{\langle a, a \rangle}$, where $\langle a, b \rangle = -\frac{1}{2}\mathrm{Trace}(ab)$ (the Killing form of $\mathfrak{so}(\mathbf{V})$). If $|a| = 1$, the observable corresponding to $\mathbf{v}_{a,\mathbf{x}_0}$ is known as the component of the *angular momentum* relative to L and the orientation of $\mathbb{M}/L$ generated by the flow. Note that the assignment

$$\mathfrak{so}(\mathbf{V}) \ni a \mapsto \mathbf{v}_{a,\mathbf{x}_0} \tag{2.6}$$

is a Lie algebra homomorphism, for each fixed $\mathbf{x}_0$.

Classical mechanical systems are not suitable models for the microworld, where one encounters phenomena such as *discreteness* of the spectra (sets of values) of some observables, and *indeterminism*, i.e., impossibility of making predictions other than in terms of probabilities. This is why one uses *quantum systems*, each of which is based on an infinite-dimensional complex Hilbert space $\mathcal{X}$, in the sense that the system's states are lines through 0 in $\mathcal{X}$, while the observables are specific, densely defined, self-adjoint operators B in $\mathcal{X}$. Thy physical spectrum of B then is just its operator spectrum, and the statistical distribution of values assumed by B in a state $\mathbb{C}\psi$, $\psi \in \mathrm{dom}(B) \smallsetminus \{0\}$, is the Borel probability measure on $\mathbb{R}$ corresponding to B and ψ via the spectral theorem. The evolution of quantum states with time is governed by Schrödinger's equation (see Remark 4.8.4).

Such a quantum system is expected to determine its *classical limit*, i.e., macroscopic (distant-view) approximation, which is a classical mechanical system with some configuration space $\mathbb{M}$. This limit operation is required to obey the *correspondence principle*, stating that the classical system's evolution and relations involving classical observables should resemble their quantum-system counterparts, with sums and products of functions $T^*\mathbb{M} \to \mathbb{R}$ corresponding to sums and composites of operators in $\mathcal{X}$. Problems with domains, as well as lack of commutativity for operators, indicate that a quantum system having a prescribed classical limit, known as a *quantization* of the latter, is in general far from unique. A natural class of quantizations can be described as follows.

A *canonical quantization* of the given classical mechanical system is obtained by using a complex vector bundle ζ, with a Hermitian fibre metric and a compatible connection $\overset{\circ}{\nabla}$, over the system's configuration space $\mathbb{M}$, along with a fixed smooth positive measure on $\mathbb{M}$. The Hilbert space $\mathcal{X}$ then consists of all L^2 sections ψ of ζ. The operators in $\mathcal{X}$ corresponding to the classical position and momentum-component observables f, $\mathbf{v}$ are $\psi \mapsto f\psi$ and $-i\hbar\overset{\circ}{\nabla}_{\mathbf{v}}$, where $\hbar$ is Planck's constant divided by 2π, needed for the agreement of units. Note that the latter operator is symmetric if $\mathrm{div}\,\mathbf{v} = 0$ relative to the measure on $\mathbb{M}$, which is rather a mild restriction on $\mathbf{v}$. However, self-adjointness of $-i\hbar\overset{\circ}{\nabla}_{\mathbf{v}}$ on a suitable domain has to be discussed on a case-to-case basis.

About the role of the Lie derivative $\mathbb{L}_{\mathbf{v}}$, as opposed to $\overset{\circ}{\nabla}_{\mathbf{v}}$, see (2.11).

References: Mackey 1963; Emch 1984; Sudbery 1986; Schweber 1962; Derdzinski 1991b.

2.6 Nonrelativistic Spin

The spin of an elementary-particle species was, so far, defined formally (§2.4) as the spin of the corresponding natural bundle ζ over the given Riemannian 3-manifold $(\mathbb{M}, g)$ (or, of its associated single or double valued representation $\mathrm{O}(T_{\mathbf{x}}\mathbb{M}) \to \mathrm{End}\,\zeta_{\mathbf{x}}$ at any $\mathbf{x} \in \mathbb{M}$). Geometrically, it can be interpreted as follows. In all relevant cases, ζ carries a natural fibre metric. The states of a quantum model of the given particle (§2.5) are lines through 0 in the Hilbert space $\mathcal{X}$ of all L^2 sections of ζ, usually represented by unit elements of $\mathcal{X}$. Let us now imagine that the particle is fully concentrated at a fixed point $\mathbf{x} \in \mathbb{M}$. Such a configuration corresponds to a Dirac delta-like distributional section of ζ, rather than an element of $\mathcal{X}$, so that it only represents an ideal, limiting case of genuine physical states. (For the particle in state $\mathbb{C}\psi$, where $\psi \in \mathcal{X}$ is L^2-unit, the function $|\psi|^2$ on $\mathbb{M}$ is the *position probability density* of the particle.) If we now perform a full rotation in the Euclidean 3-space $T_x\mathbb{M}$, about any fixed axis, then, by (2.5), the "internal state" of the particle in the fibre ζ_x *will undergo s "full rotations", where s is the spin* of the particle.

In particular, rotating a localized fermion by 360° around an axis, one ends up with a configuration different from the initial one.

Reference: Schweber 1962.

2.7 Spin and Angular Momentum

As a result of the canonical quantization (§2.5), if $\mathbb{M}$ is an affine Euclidean 3-space, $x_0 \in \mathbb{M}$ and $a \in \mathfrak{so}(\mathbf{V})$, the operator

$$\ell^a_{x_0} = -i\hbar \overset{\circ}{\nabla}_{v_{a,x_0}} \tag{2.7}$$

is the *quantized a-component of the angular momentum* about x_0 (Example 2.5.1(b)). If $|a| = 1$, $\ell^a_{x_0}$ may also be called the (quantized) angular momentum component about the axis $L = x_0 + \operatorname{Ker} a$, corresponding to the orientation of the plane $\mathbb{M}/L$ determined by a as in Example 2.5.1(b). If $\overset{\circ}{\nabla}$ is flat, for each fixed x_0 the assignment

$$\mathfrak{so}(\mathbf{V}) \ni a \mapsto -\frac{i}{\hbar}\ell^a_{x_0} \tag{2.8}$$

is a Lie algebra homomorphism. More generally, (2.8) is homomorphic up to a curvature term. However, in all important cases ζ is a natural bundle over $\mathbb{M}$ with the canonical connection $\overset{\circ}{\nabla}$ (§1.18), which is flat since so is $\mathbb{M}$.

In the case where $(\mathbb{M}, g)$ is a Riemannian manifold of any dimension and ζ is a natural bundle over $(\mathbb{M}, g)$ (defined similarly as in §1.18), the single or double valued group homomorphisms $O(T_x\mathbb{M}) \to \operatorname{Aut}\zeta_x$, $x \in \mathbb{M}$ (cf. §1.18) lead to a family of Lie algebra homomorphisms $\tau_x : \mathfrak{so}(T_x\mathbb{M}) \to \operatorname{End}\zeta_x$. For any section φ of the bundle $\mathfrak{so}(T\mathbb{M})$ of Lie algebras, formula $s^\varphi(x) = i\hbar\tau_x(\varphi(x))$ defines a section s^φ of $\operatorname{End}\zeta$. Obviously, $\varphi \mapsto -\frac{i}{\hbar}s^\varphi$ is a homomorphism of Lie algebras, with brackets defined pointwise. If, in addition, $\mathbb{M}$ is a Euclidean affine 3-space with the translation space $\mathbf{V}$, each $a \in \mathfrak{so}(\mathbf{V})$, regarded as a constant (parallel) section of $\mathfrak{so}(T\mathbb{M}) = \mathbb{M} \times \mathfrak{so}(\mathbf{V})$ gives rise to the quantum observable s^a (which is also *bounded*, as a self-adjoint operator), known as the *a-component of the spin*. If $|a| = 1$, s^a is also referred to as the spin component relative to the direction $\operatorname{Ker} a$ in $\mathbb{M}$ (i.e., the line $\operatorname{Ker} a$ in $\mathbf{V}$) and the orientation of $\mathbf{V}/\operatorname{Ker} a$ determined by a. The fact that

$$\mathfrak{so}(\mathbf{V}) \ni a \mapsto -\frac{i}{\hbar} s^a \tag{2.9}$$

is a Lie algebra homomorphism, indicates that there is some similarity between s^a and $\ell^a_{\mathbf{x}_0}$ (cf. (2.8)), enhanced by the fact that both s^a and $\ell^a_{\mathbf{x}_0}$ represent the same physical units of angular momentum. It is therefore natural to think of s^a as some sort of angular momentum components. However, since s^a for a with $|a| = 1$ depends only on the *direction* Ker a in $\mathbb{M}$ rather than the specific line $L = \mathbf{x}_0 + \text{Ker } a$ (as was the case with $\ell^a_{\mathbf{x}_0}$), one should regard s^a as components of the *intrinsic angular momentum* of the particle in question, as opposed to the $\ell^a_{\mathbf{x}_0}$, which are the components of the *orbital angular momentum* about the rotation axis L. The idea of a particle spinning about its own axis is a useful analogy.

The fact that s^a is a $\overset{\circ}{\nabla}$-parallel section of $\text{End}\,\zeta$ (as a was parallel in $\mathfrak{so}(T\mathbb{M})$ means that $\ell^a_{\mathbf{x}_0}$ and s^b commute for any $a, b \in \mathfrak{so}(\mathbf{V})$ and $\mathbf{x}_0 \in \mathbb{M}$. Consequently,

$$\mathfrak{so}(\mathbf{V}) \ni a \mapsto -\frac{i}{\hbar}(\ell^a_{\mathbf{x}_0} + s^a) \tag{2.10}$$

is a Lie algebra homomorphism, while no other nonzero combination of $\ell^a_{\mathbf{x}_0}$ and s^a has this property. In view of the above interpretations of $\ell^a_{\mathbf{x}_0}$ and s^a, it is therefore natural to call the quantum observable $\ell^a_{\mathbf{x}_0} + s^a$ the a-component of the *total angular momentum* about the axis $L = \mathbf{x}_0 + \text{Ker } a$ (if $a \neq 0$). One can verify that

$$\ell^a_{\mathbf{x}_0} + s^a = -i\hbar \mathbb{L}_{\mathbf{v}_{a,\mathbf{x}_0}} \tag{2.11}$$

$\mathbb{L}$ being the Lie derivative, so that $\ell^a_{\mathbf{x}_0} + s^a$ is a natural object to discuss, both for physical and geometric reasons.

If $|a| = 1$, the operators $\ell^a_{\mathbf{x}_0}$ and s^a have pure point spectra with the following eigenvalues:

$$k\hbar, \quad k \in \mathbb{Z}$$

for $\ell^a_{\mathbf{x}_0}$, and

$$j\hbar, \quad j = -s, -s+1, \ldots, s-1, s \tag{2.12}$$

for s^a in the case where the natural bundle ζ over the Euclidean affine 3-space $\mathbb{M}$ is irreducible and $s \in \{0, \frac{1}{2}, 1, \frac{3}{2}, \ldots\}$ is its spin (§2.4). The eigenvalues of $\ell^a_{\mathbf{x}_0} + s^a$ now can easily be found using the fact that $\ell^a_{\mathbf{x}_0}$ and s^a commute. It is the *total* angular momentum that is conserved (in a suitable, natural sense) in all microworld processes, while its orbital and intrinsic parts in general are not. Measuring how the orbital angular momentum changes in observed interactions is one possible way of determining the spins of various elementary-particle species.

References: Mackey 1963; Emch 1984; Sudbery 1986; Schweber 1962; Derdzinski 1991b.

2.8 Parity Versus Orbital Parity

In §2.7 a family of quantum observables representing conserved physical quantities, the total angular momentum components, was obtained by value-wise addition of two Lie algebra homomorphisms whose values commute (so that the sum is also a homomorphism). In terms of Lie groups, this can be described as follows. Let $\mathbf{V}$ be the translation space of the Euclidean affine 3-space $\mathbb{M}$, as in §2.7. The group $SO(\mathbf{V}) \cong SO(T_{\mathbf{x}_0}\mathbb{M})$ acts on $\mathbb{M}$ by rotations about a fixed point $\mathbf{x}_0$. The canonical (flat) connection $\overset{\circ}{\nabla}$ in the given natural bundle ζ over $\mathbb{M}$ (§1.18) gives rise to the bundle isomorphism $\zeta \cong \mathbb{M} \times \zeta_{\mathbf{x}_0}$ obtained by $\overset{\circ}{\nabla}$-parallel displacements. Thus, the action of $SO(\mathbf{V})$ can be lifted to the total space ζ so that it is trivial on the fibre factor (and has $\zeta_{\mathbf{x}_0}$ as its fixed-point set). On the other hand, the tangential actions of $SO(\mathbf{V}) \cong SO(T_{\mathbf{x}}\mathbb{M})$ on $T_{\mathbf{x}}\mathbb{M}$ give rise, by naturality, to its (possibly double-valued) actions on all fibres $\zeta_{\mathbf{x}}$, $\mathbf{x} \in \mathbb{M}$, i.e., to an action of $SO(\mathbf{V})$ through bundle automorphisms of ζ. The resulting homomorphisms

$$SO(\mathbf{V}) \to \text{Diff}\,\zeta, \quad SO(\mathbf{V}) \xrightarrow[\text{valued}]{\substack{\text{single or} \\ \text{double}}} \text{Aut}\,\zeta \subset \text{Diff}\,\zeta$$

have commuting values and so their value-wise product is a homomorphism $SO(\mathbf{V}) \to \text{Diff}\,\zeta$, representing the total angular momentum. (It may be double-valued, but becomes single-valued when lifted to $SU(S)$ as in §2.1.)

For $\mathbb{M}, \zeta$ and $\overset{\circ}{\nabla}$ as above, the preceding discussion may be repeated with $SO(\mathbf{V})$ replaced by another normal subgroup of $O(\mathbf{V})$, namely its center $\mathbb{Z}_2 = \{\text{Id}, -\text{Id}\}$. For any $\mathbf{x}_0 \in \mathbb{M}$, the generator $-\text{Id}_{\mathbf{V}}$ of $\mathbb{Z}_2$ acts on $\mathbb{M}$ by the reflection $R_{\mathbf{x}_0}$ about $\mathbf{x}_0$, $R_{\mathbf{x}_0}(\mathbf{x}) = \mathbf{x}_0 + (\mathbf{x}_0 - \mathbf{x})$, which gives rise to an involutive diffeomorphism $\widetilde{R}_{\mathbf{x}_0}$ of $\zeta \cong \mathbb{M} \times \zeta_{\mathbf{x}_0}$, operating trivially on the fibre factor. On the other hand, since ζ is natural, the differential $-\text{Id}_{T_{\mathbf{x}}\mathbb{M}}$ of $R_{\mathbf{x}}$ at any $\mathbf{x} \in \mathbb{M}$ acts on $\zeta_{\mathbf{x}}$, possibly in a double-valued fashion. Let us now assume that the spins of the irreducible summands of ζ are all integral, or all non-integral (so that ζ is either "bosonic", or "fermionic"). The action of $-\text{Id}_{T_{\mathbf{x}}\mathbb{M}}$ on $\zeta_{\mathbf{x}}$ then consists of a transformation $\mathbb{P}_{\mathbf{x}}$ (defined up to a sign for fermionic bundles), with $\mathbb{P}_{\mathbf{x}}^2 = \text{Id}$ (bosonic case), or $\mathbb{P}_{\mathbf{x}}^2 = -\text{Id}$ (fermionic case). In either case, the bundle automorphism $\mathbb{P}$ of ζ consisting of all the $\mathbb{P}_{\mathbf{x}}$ (or, either of its global C^{∞} branches) commutes with each $\widetilde{R}_{\mathbf{x}_0}$. We then define the *total parity operator* (quantum observable) about $\mathbf{x}_0$ by

$$\mathcal{P}_{\mathbf{x}_0} = \widetilde{R}_{\mathbf{x}_0} \circ \mathbb{P}$$

(bosonic case), or

$$\mathcal{P}_{\mathbf{x}_0} = \widetilde{R}_{\mathbf{x}_0} \circ (-i\mathbb{P})$$

(fermionic case), so that $\mathcal{P}_{\mathbf{x}_0}^2 = \text{Id}$, and the Hilbert space $\mathcal{X}$ is the direct sum of (± 1)-eigenspaces of $\mathcal{P}_{\mathbf{x}_0}$. In the fermionic case, $\mathcal{P}_{\mathbf{x}_0}$ is only defined up to a sign, but it still makes sense to *compare* parities of any two of its eigenstates. Thus, the total parity $\mathcal{P}_{\mathbf{x}_0}$ about $\mathbf{x}_0$ is the product of the orbital parity $\widetilde{R}_{\mathbf{x}_0}$ and the *intrinsic* parity $\mathbb{P}$ or $-i\mathbb{P}$ independent of the reference point $\mathbf{x}_0$.

More generally, the total parity of a system of particles is the product of their (intrinsic) parities and the orbital parity of their configuration about their center of mass, assuming, of course, that the state in question is a parity eigenstate. The total parity thus defined is conserved in strong and electromagnetic interactions (but not always in weak interactions), which makes it possible to determine parities of many particle species from various observed microworld processes.

Reference: Ryder 1986.

2.9 Isospin in Nature

The lightest hadrons known to exist in nature are listed in the tables of §1.15, where each particle species is represented by a dash ——, plotted against the corresponding mass, spin and parity.

Evidently, most hadrons occur in clusters, known as *multiplets* (quadruplets, triplets, doublets); those which do not are said to form *singlets*. The members of any given multiplet have the same spin and parity and their masses are *almost* equal, while their electric charges, in units of proton charge, form a progression of step 1. The mass differences in all multiplets are exaggerated, to make the diagrams in §1.15 legible. Actually, the relative mass difference $(m_{\max} - m_{\min})/m_{\min}$ in each multiplet is less than 1% except for the *pion triplet* π, where it stays within 3.5%.

The symbol N represents the *nucleon doublet*, consisting of the *proton* $p = N^+$ and *neutron* $n = N^0$, the superscripts standing, as usual, for the electric charge.

The antiparticles of all members of any given multiplet X also form a multiplet, denoted $\overline{X}$. For baryon multiplets X, $\overline{X} \neq X$, since the antiparticle formation changes the parity of fermions (§§1.14, 2.2, 2.4). This is not necessarily the case for meson multiplets, some of which coincide with their antimultiplets. For instance, $\overline{\pi} = \pi$.

Let a hadron multiplet $X = \{X_1, \ldots, X_k\}$ be ordered by increasing electric charges Q_j, so that $Q_j = Q_1 + j - 1$, $j = 1, \ldots, k$. The *isospin* I of the multiplet X (and of each member particle X_j) is, by definition, the number

$$I = \frac{k-1}{2} \in \{0, \frac{1}{2}, 1, \frac{3}{2}, \ldots\}. \tag{2.13}$$

One also defines the *isospin component* I_3 of each particular particle species X_j, $j = 1, \ldots, k$, by

$$I_3 = j - \frac{k+1}{2} \in \{-I, -I+1, \ldots, I-1, I\}. \tag{2.14}$$

Thus, in contrast with I, I_3 *does* depend on j. The subscript 3 in I_3 is due to its traditionally being referred to as the "isospin component with respect to the *third* coordinate axis in isospin space", cf. §2.10. Consequently, $X_1, \ldots, X_n$ have I_3 equal to $-I, -I+1, \ldots, I-1, I$, respectively. The *hypercharge* Y of each X_j (or, of the multiplet X) is defined to be twice the average electric charge in X,

$$Y = \frac{2}{k} \sum_{j=1}^{n} Q_j = 2Q_1 + k - 1, \tag{2.15}$$

so that *every* hadron satisfies the relation

$$Q = I_3 + \frac{1}{2}Y, \tag{2.16}$$

Q being its electric charge. (See, however, Remark 8.1.1.)

References: Ryder 1986; Sudbery 1986.

2.10 Geometry of Isospin

The most characteristic property of hadron multiplets (§2.9) seems to lie in the fact that, apart from the differences in the electric charge, members of any given multiplet appear essentially identical. Regarding the slight differences in masses among the multiplet members as "caused" by charge differences is also suggested by the fact that mass differences between pairs of nuclei with equal numbers of nucleons are correctly accounted for by the *electromagnetic* contribution to their binding energies. It is therefore natural to consider a common generalization (§1.8) of the particle species X_j, $j = 1, \ldots, k$, forming a given hadron multiplet X. More precisely, one obtains it as a special case of generalized spin symmetry (§1.9), the specific form of this symmetry, which we now proceed to describe, being justified by its analogy with nonrelativistic spin (discussed later in §2.11).

All particle species X_j, $j = 1, \ldots, k$, live in the same (natural) vector bundle η over $(\mathcal{U}, g)$, as they have the same spin and parity. Here $(\mathcal{U}, g)$ may stand for a spacetime or a Riemannian 3-manifold, as in §1.18, depending on whether the desired discussion is to be relativistic or not. For a fixed Pauli triple $(S, \Omega, \langle, \rangle)$ (§2.1), common to all hadron multiplets, let $\imath^{(1/2)} = \mathcal{U} \times S$ be the corresponding *isospinor bundle*, regarded as endowed with the "constant" fibrewise structures coming from Ω and $\langle, \rangle$. If $I = \frac{k-1}{2}$ is the isospin of the multiplet X, we define the *isospin I bundle* $\imath^{(I)}$ over $\mathcal{U}$ by $\imath^{(I)} = \mathcal{U} \times (\Sigma_I(T) \otimes \mathbb{C})$ for $I \in \mathbb{Z}$, with T as in §2.1 (cf. §2.3), and $\imath^{(I)} = \mathcal{U} \times \Sigma_I(S)$ for $I \notin \mathbb{Z}$. (Equivalently, we may set $\imath^{(I)} = \mathcal{U} \times S^{(I)}$, where, for any I, $S^{(I)}$ is the $(2I)$th symmetric power of S. See Example 2.3.7.)

As a generalized particle, the multiplet X is represented by the bundle $\imath^{(I)} \otimes \eta$. Let a be any unit element of $\mathfrak{so}(\mathcal{T})$, regarded as a constant section of $\operatorname{End} \imath^{(1)}$ and also acting naturally on each $\imath^{(I)}$ (cf. §2.3). (Since the Euclidean 3-space $\mathcal{T}$ is oriented, a may also be identified with a unit vector in $\mathcal{T}$, i.e., a constant unit real section of $\imath^{(1)}$.) The a-*isospin component operator* is the bundle endomorphism

$$I_a = -ia \otimes \operatorname{Id}_\eta \tag{2.17}$$

of $\imath^{(I)} \otimes \eta$, also to be thought of, in the nonrelativistic case, as a self-adjoint bounded operator (observable) in the space of all L^2 sections of $\imath^{(I)} \otimes \eta$ (§2.5). In each fibre, I_a has the eigenvalues

$$-I, -I+1, \ldots, I-1, I,$$

all simple. As in §1.9, we now obtain the splitting

$$\imath^{(I)} \otimes \eta = \eta_{-I} \oplus \eta_{-I+1} \oplus \ldots \oplus \eta_{I-1} \oplus \eta_I \tag{2.18}$$

the summands η_α being constant subbundles of $\imath^{(I)} \otimes \eta$ with $\eta_\alpha = \operatorname{Ker}(I_a - \alpha \cdot \operatorname{Id}_{\imath^{(I)} \otimes \eta}) = \iota_\alpha \otimes \eta$, while $\iota_\alpha = \operatorname{Ker}(-ia - \alpha \cdot \operatorname{Id}_{\imath^{(I)}})$. Thus, each η_α can be identified with η by choosing a constant unit section of the product line bundle ι_α. In this way, each choice of a unit constant section a of $\mathfrak{so}(\mathcal{U} \times \mathcal{T})$ gives rise to $k = 2I+1$ particle species represented by the η_α, $\alpha = -I, -I+1, \ldots, I$. However, the isospin symmetry is "broken" (imperfect), since nature has selected a *distinguished* a, reflecting the presence of the electromagnetic interaction. For this specific a, each η_α, $\alpha = -I, -I + 1, \ldots, I$, represents the multiplet member whose isospin component I_3 equals α.

Assuming that $\imath^{(1/2)}$, and hence all $\imath^{(I)}$, are *product bundles*, i.e., carry a distinguished flat connection with trivial holonomy, is not absolutely necessary. One could have started from a bundle $\imath^{(1/2)}$ over $\mathcal{U}$ with an SU(2) structure (i.e., objects $\Omega, \langle, \rangle$ forming a Pauli triple in each fibre), and then obtain all $\imath^{(I)}$ by the same functorial construction as above (or even simpler, using Example 2.3.7). Splittings of $\imath^{(I)} \otimes \eta$ then would be parametrized by all unit sections a of $\mathfrak{so}(\operatorname{Re} \imath^{(1)})$ ($\operatorname{Re} \imath^{(1)}$ being the real form of $\imath^{(1)}$), rather than just by "constant" ones, and electromagnetism would specify one such section. However, physically, there seems to be no need of representing isospin by this huge, infinite-dimensional symmetry.

References: Sudbery 1986; Ryder 1986.

2.11 Isospin Versus Nonrelativistic Spin

A useful analogy between isospin and nonrelativistic spin (§§2.10, 2.7) can be described as follows. Let a particle of spin s live in an irreducible natural bundle ζ over a Euclidean affine 3-space $\mathbb{M}$. For simplicity, let us fix an orientation of $\mathbb{M}$, thus ignoring parity, so that the lifting functor of ζ is restricted to orientation-preserving linear transformations between tangent spaces. Clearly (see §2.4), ζ then may be identified with $\imath^{(s)}$ defined as in §2.10, with $\imath^{(1/2)}$ equal to a fixed Pauli spinor bundle over $\mathbb{M}$, so that, as $\mathbb{M}$ is Euclidean, $\imath^{(1/2)} = \mathbb{M} \times S$, $(S, \Omega, \langle, \rangle)$ being a suitable Pauli triple. (We exclude here strictly neutral particles. Thus, all vector bundles involved are complex.) Any unit element $a \in \mathfrak{so}(\mathbf{V})$, where $\mathbf{V}$ is the translation space of $\mathbb{M}$, i.e., any unit constant section a of $\mathfrak{so}(\operatorname{Re}\imath^{(1)})$, now leads to the constant bundle endomorphism $I_a = -\frac{1}{\hbar}s^a$ (§2.7) with the eigenvalues $-s, -s + 1, \ldots, s - 1, s$ (see (2.12)), and so it gives rise to the splitting

$$\zeta = \imath^{(s)} = \imath^{(s)} \otimes (\mathbb{M} \times \mathbb{C}) = \eta_{-s} \oplus \eta_{-s+1} \oplus \ldots \oplus \eta_s$$

into the α-eigenvalue bundles η_α of I_a, $\alpha = -s, -s + 1 \ldots, s$, each of which is a product line bundle and can be identified with $\mathbb{M} \times \mathbb{C}$ by choosing a constant unit section. Thus, formally the situation is quite similar to the case of isospin in §2.10 with $I, \imath^{(I)}$ and η now replaced by s, $\zeta = \imath^{(s)}$ and $\mathbb{M} \times \mathbb{C}$, respectively. In other words, any choice of a as above, which, for charged particles, amounts to switching on a constant magnetic field pointing along the rotation axis of a, makes our particle of spin s appear as a generalization of $2s + 1$ different particle species, each of which has spin 0, i.e., lives in $\mathbb{M} \times \mathbb{C}$. This is actually what happens when a beam of charged particles (which may even be nuclei) passes through a magnetic field as above: it becomes split into $2s + 1$ subbeams. The difference between this phenomenon and the isospin symmetry that justifies regarding the particle of spin s as a single species rather than a generalization of $2s + 1$ scalar, i.e., structureless particles (§1.2), lies in the fact that, in contrast with isospin, the spin symmetry is *exact* ("perfect") since no space direction is physically distinguished by the presence of a constant magnetic field, or otherwise.

Reference: Sudbery 1986.

2.12 *G*-parity

Let us consider a hadron multiplet $X = \{X_{-I}, X_{-I+1}, \ldots, X_I\}$ of isospin I, ordered by increasing electric charge, so that the isospin component I_3 for the particle species X_α equals α, $\alpha = -I, -I + 1, \ldots, I$ (§2.9). Then, X coincides with its antimultiplet $\overline{X}$ if and only if it contains a strictly neutral particle (§1.6). Also, $X = \overline{X}$ implies that X consists of mesons (i.e., bosons, of integral spin) of hypercharge $Y = 0$, so that the electric charge $Q = I_3$ for each X_α, while $I \in \mathbb{Z}$ and X_0 is strictly neutral. All of this follows from the fact that electric charges in X form a progression and are reversed under antiparticle formation, while $Q = I_3 + \frac{1}{2}Y$ by (2.16), and two multiplets have to be either identical or disjoint, the latter being always that case for baryon (fermion) multiplets as antiparticle formation changes their parity (§1.14).

Let us assume, throughout this section, that the given multiplet X is "strictly neutral" ($\overline{X} = X$). Whether our description is relativistic or not, all electrically charged member particles X_α of X, $\alpha \neq 0$, live in the complexification $\zeta \otimes \mathbb{C}$ of a suitable real vector bundle ζ, while the neutral particle X_0 is represented by ζ itself (see §1.7, §2.4). This contradicts the general requirement that all members of any fixed multiplet correspond, essentially, to the same bundle (cf. §2.10). Consistency of our approach can be easily restored by regarding states of X_0 as sections $\zeta \otimes \mathbb{C}$ while keeping in mind that they actually have to be valued in one of the real subbundles $\zeta, i\zeta$ depending on whether the C-parity of X_0 is 1 or -1 (§1.7).

As $I \in \mathbb{Z}$, both $\zeta \otimes \mathbb{C}$ and $\imath^{(I)}$ (§2.10) are given as complexifications of real bundles, and hence so is their tensor product $\imath^{(I)} \otimes \zeta \otimes \mathbb{C}$ (which may also be described as $\imath^{(I)} \otimes \zeta$). This gives rise to the *complex conjugation antiautomorphism* ( )$^-$ of $\imath^{(I)} \otimes \zeta \otimes \mathbb{C}$. We would like to extend the discussion in §1.7, which led to the notion of C-parity for strictly neutral particles, to the bundle $\imath^{(I)} \otimes \zeta \otimes \mathbb{C}$ representing a generalized particle corresponding to the multiplet $X = \overline{X}$. However, the complex conjugation cannot play here the same role as in §1.7. In fact, let a denote, here and in the sequel, the fixed unit element of $\mathfrak{so}(T)$, i.e., a constant unit section of $\mathfrak{so}(\mathrm{Re}\,\imath^{(1)})$, "chosen" by nature to represent the electromagnetic interaction. Since I_a in (2.17) is imaginary (anticommutes with ( )$^-$), ( )$^-$ sends its α-eigenspace subbundle η_α in $\imath^{(I)} \otimes \zeta \otimes \mathbb{C}$ onto $\eta_{-\alpha}$, $\alpha = -I, -I+1, \ldots, I$. As η_α stands for the member particle X_α, one cannot think of ( )$^-$ as keeping fixed all states in a reasonably large class, for such states would not describe the individual species X_α with $\alpha \neq 0$. A natural way out of this difficulty might consist in replacing ( )$^-$ by the complex conjugation corresponding to some other, relatively natural real form β of $\imath^{(I)} \otimes \zeta \otimes \mathbb{C}$, that is, a real subbundle β with $\imath^{(I)} \otimes \zeta \otimes \mathbb{C} = \beta \oplus i\beta$.

To this end, let us consider any fixed line L through 0 in $(\mathrm{Ker}\,a)^\perp \subset T$ (§2.10), so that L is orthogonal to the rotation axis of a. The reflection in L (rotation by 180° about L) is an element of $\mathrm{SO}(T)$ and so it operates, naturally, as a real automorphism R_L of $\Sigma_I(T) \otimes \mathbb{C}$, also regarded as a constant bundle automorphism of $\imath^{(I)} = \mathcal{U} \times (\Sigma_I(T) \otimes \mathbb{C})$ (§2.10) or $\imath^{(I)} \otimes \zeta \otimes \mathbb{C}$ (after tensoring by Id), satisfying $R_L^2 = \mathrm{Id}$, commuting with ( )$^-$ and anticommuting with I_a. The composite $R_L \circ (\ )^- = (\ )^- \circ R_L$ is a good candidate for a "new" complex conjugation in $\imath^{(I)} \otimes \zeta \otimes \mathbb{C}$, since it is a constant antilinear bundle involution, leading to the "real form" $\beta_L = \{\psi : R_L\overline{\psi} = \psi\}$, and *commuting* with I_a, i.e., leaving each subbundle η_α

invariant. However, $R_L \circ (\ )^-$ obviously *does* depend on the choice of the line $L \subset (\operatorname{Ker} a)^\perp$. For a fixed L, proceeding as in §1.7, we may expect that states of the generalized particle actually live either in the real subbundle β_L, or in $i\beta_L$. We then say that the *G-parity* ϵ_G of the multiplet X (or, of each member particle X_α) equals 1 or, respectively, -1. In other words, all "physical" states of the generalized particle are eigenstates of $R_L \circ (\ )^-$ for the eigenvalue ϵ_G.

It turns out that postulating a *conservation law* for G-parity in strong interactions (to be discussed in §2.14) one can determine, using experimental data, a consistent collection of values of G-parity for most known hadrons that belong to the same multiplet X as their antiparticles (so that $\overline{X} = X$). It is therefore natural to assume that, besides the unit element a of $\mathfrak{so}(T)$ "responsible" for electromagnetism, nature has also selected a line L through 0 in the isospin space T, perpendicular to $(\operatorname{Ker} a)^\perp$, which accounts for the G-parity known from experiment.

The G-parity ϵ_G of each particle X_α in a given mulitplet X with $\overline{X} = X$ is related to the C-parity ϵ_C of the neutral member X_0 (§1.7) by

$$\epsilon_G = (-1)^I \epsilon_C \tag{2.19}$$

where I is the isospin of X (and so $I \in \mathbb{Z}$). In fact, R_L acts on the kernel of I_a via multiplication by $(-1)^I$. To see this, note that a spherical harmonic of degree I in $\mathbb{R}^3$ with Cartesian coordinates x, y, z, invariant under rotations about the x-axis (to be thought of as the group $t \mapsto \exp ta$) must be a (suitable) polynomial in x and $y^2 + z^2$ and so the highest power of x in each of its monomials is congruent to I mod 2. Composing the spherical harmonic with the map $(x, y, z) \mapsto (-x, y, -z)$, corresponding to R_L, therefore results in its multiplication by $(-1)^I$.

Reference: Ryder 1986.

2.13 Strangeness and Other Generalized Charges

A *generalized charge* is a function assigning a real number to each particle species of a given class in such a way that its sign changes under antiparticle formation. Among the most obvious examples, we have the *electric charge* (see §4.14), defined for all particles, and the *parity* restricted to the class of fermions, cf. §1.14. (Usually, only those generalized charges are of interest which are *conserved* in the sense explained in §2.14.) Further examples are the *electron number* L_e, *muon number* L_μ, *tauon number* L_τ, *baryon number* B, as well as the *lepton number* $L = L_e + L_\mu + L_\tau$. They are defined for all particles, in such a way that $L_e = 1$ for the electron e$^-$ and the electronic neutrino ν_e, $L_e = -1$ for their antiparticles, and $L_e = 0$ for all other particles. The definitions of L_μ and L_τ are similar, with the pair e$^-$, ν_e

simply replaced by μ^-, ν_μ or τ^-, ν_τ. Thus, the lepton number L equals 1 for leptons proper, -1 for antileptons and 0 for all other particles (§1.14). As for the baryon number B, it is ± 1 for baryons and 0 for all remaining particles. It also turns out that mutually opposite values of B can be selected for each baryonic particle-antiparticle pair so as to make B a conserved quantity, uniquely determined by requiring that $B = 1$ for the proton p. For instance, $B = 1$ for the neutron n in view of the β decay n $\to$ p$+$e$+\bar{\nu}_e$. Consequently, it makes sense to speak of *antibaryons*, with $B = -1$, as opposed to *baryons proper* with $B = 1$.

Other important generalized charges, defined for all hadrons, are the *isospin component* I_3 and *hypercharge* Y (§2.9). Finally, the *strangeness* S is a generalized charge introduced by Gell-Mann and Nishijima in 1953 to account for reactions such as

$$\pi^- + p \to \Lambda^0 + K^0, \quad \text{or} \quad \pi^- + p \to \Sigma^- + K^+,$$

due to the strong interaction, which are immediately followed by the *weak* decays

$$\Lambda^0 \to p + \pi^-, \qquad K^0 \to \pi^+ + \pi^-,$$

or

$$\Sigma^- \to n + \pi^-, \qquad K^+ \to \pi^+ + \pi^0.$$

Defining the $\mathbb{Z}$-valued generalized charge S (for hadrons) to be the same for all members of any multiplet, with the values 1 for K, -1 for Λ and Σ, 0 for $N = \{n, p\}$ and π, etc., one can "explain" the above reactions as follows. The strangeness S is conserved in strong interactions, and so "strange"' particles such as $\Lambda^0, K^0, \Sigma^-, K^+$ can only be produced via strong interaction *in pairs* (out of "nonstrange" hadrons), so that the values of S in the pair sum up to 0. (It is the pairs-only, or *associated* production, that people found strange at the time.) However, S need not be conserved in weak interactions, and so strange particles may decay *singly*, as long as the decay is weak. Using experimental data as well as the *Gell-Mann–Nishijima relation*

$$Y = B + S \tag{2.20}$$

i.e., by (2.16),

$$Q = I_3 + \frac{1}{2}(B + S), \tag{2.21}$$

valid for the electric charge Q, isospin component I_3, hypercharge Y, baryon number B and strangeness S of the *lightest* hadrons, values of S were found for all known hadrons. Also, the value of S discovered for a partially known multiplet was in some cases used to successfully predict the existence of further multiplet members, as the electric charges occurring in a given multiplet form a progression, symmetric about $\frac{1}{2}Y = \frac{1}{2}(B + S)$.

Further, interesting generalized charges defined for hadrons and found using both theoretical arguments and experimental data involving heavier hadrons are the *charm C*, *bottomness* (or: *beauty*) *b* and *topness* (also known as *truth*), *t*. They are all zero for the lightest hadrons, listed in §1.15; see, however, §8.1. To be valid for *all* known hadrons, the Gell-Mann–Nishijima relation (2.21) has to be modified as follows:

$$Q = I_3 + \frac{1}{2}(B + S + C + b + t). \qquad (2.22)$$

See also Remark 8.1.4.

References: Ryder 1986; Sudbery 1986.

2.14 A Summary of Particle Invariants

A *particle invariant* is a function assigning to each particle species of a given class an element of a specific Abelian semigroup (often, it is an Abelian group). Such an invariant is referred to as *additive* or *multiplicative* depending on what notational convention is used for the (semi)group operation.

The semigroups in which invariants most frequently used in particle physics are valued are:

$$[0, \infty), \quad (0, \infty], \quad \mathbb{R} = (-\infty, \infty), \quad \mathbb{Z}, \quad \mathbb{Z}_+ = \{0, 1, 2, \ldots\},$$
$$\frac{1}{2}\mathbb{Z} = \{y \in \mathbb{R} : 2y \in \mathbb{Z}\}, \quad \frac{1}{2}\mathbb{Z}_+,$$

all regarded as additive subsemigroups of $\mathbb{R}$, and, in addition, $\mathbb{Z}_2$, usually represented by the multiplicative group $\{1, -1\}$. For further examples, see the end of this section.

The invariants taking values in the first three of the semigroups just listed are, respectively, the *mass m*, *mean life T*, and the *magnetic moment*. (Although only a finite number of particle species are known, there seems to be no valid reason to expect these values to lie in smaller semigroups.) Note that $m > 0$ except for the *massless particles*, which are the neutrinos ν_e, ν_μ, ν_τ, antineutrinos $\overline{\nu}_e, \overline{\nu}_\mu, \overline{\nu}_\tau$, photon γ and eight species of gluons. Also, the mean life T (the average time needed for the particle to decay spontaneously, without external interaction) is infinite for the *stable particles*: $e^-, e^+, \nu_e, \nu_\mu, \nu_\tau, \overline{\nu}_e, \overline{\nu}_\mu, \overline{\nu}_\tau, \gamma$, as well as the proton p and antiproton $\overline{p}$, while $T < \infty$ for all other particles (with T equal to approx. 15 minutes for the neutron n and antineutron $\overline{n}$, and $T < 2.2 \times 10^{-6}$ seconds for everything else). We do not discuss here the mean life of quarks and gluons, as they cannot exist in isolation.

Invariants valued in the remaining semigroups $\mathbb{Z}, \mathbb{Z}_+, \frac{1}{2}\mathbb{Z}, \frac{1}{2}\mathbb{Z}_+$ and $\mathbb{Z}_2 = \{1, -1\}$ are called *discrete invariants* or *quantum numbers*.

An additive (multiplicative) particle invariant is said to be *conserved* in a given class of interactions if for any collection of particles for which the invariant is defined, the sum (product) of the values the invariant assigns to them remains constant in time, even after interactions of the given type, which may affect the identities and/or the number of particles. (In such a collection, a given species may occur more than once, and then the corresponding value enters the sum/product with appropriate multiplicity/power.) The resulting sum/product then consistently defines the value of the invariant for *composite objects* whose constituents are held together by the given interaction ("force"). Thus, invariants conserved in strong interactions and defined for nucleons will also make sense for *atomic nuclei*, while those defined both for electrons and nuclei and conserved in electromagnetic interactions can be naturally extended to *atoms*.

Of the additive generalized charges (which are a special type of quantum numbers, defined in §2.13), Q, L_e, L_μ, L_τ, L and B are conserved in *all* interactions, while I_3, Y, S, C, b and t are conserved in strong and electromagnetic interactions.

The mass m, mean life T, magnetic moment, spin s and isospin I are *not* conserved, as additive invariants, in any reasonably large class of interactions. However, the mass m and spin s can naturally be regarded as *intrinsic aspects* of some universally conserved *dynamical* (motion-related) quantities, namely, the 4-momentum (energy-momentum) and, respectively, angular momentum (§4.3, §2.7). In particular, spins satisfy the *vector addition rule*: The spin of a composite object obtained by putting together k particles with spins $s_1, \ldots, s_k$ and imparting to them ℓ units of orbital angular momentum about their center of mass (§1.13), can be any number s with

$$\sum_{j=0}^{k} s_j \geq s \geq \max(0, 2 \max_{0 \leq j \leq k} s_j - \sum_{j=0}^{k} s_j), \quad \sum_{0 \leq j \leq k} s_j - s \in \mathbb{Z}, \qquad (2.23)$$

where we have set $s_0 = \ell$. This leads to finitely many possible values for s; see §3.2 for details. A similar vector addition rule holds, for the same reason, for the isospin I, however, without any "orbital" contribution, and only in the case where the constituents are held together by the strong interaction. (Sometimes this is called the "isospin conservation law".)

The parity ϵ, C-parity ϵ_C and G-parity ϵ_G are multiplicative quantum numbers defined, respectively, for all particles other than leptons, or, all strictly neutral particles, or, all mesons belonging to strictly neutral multiplets (see §§2.4, 1.7, 2.12). Of these, ϵ_C is conserved in strong and electromagnetic interactions, while ϵ_G is conserved in strong interactions. As for ϵ, it resembles the spin in being the intrinsic aspect of something "quasi-dynamical" (position-related, namely, the orbital parity, cf. §2.8) and consequently, it satisfies a *modified* conservation law ("multiplication rule"). Specifically, the composite object formed using k particles of parities $\epsilon_1, \ldots, \epsilon_k$ with ℓ units of orbital angular momentum about their common center of mass has the parity

$$\epsilon = (-1)^{\ell} \epsilon_1 \ldots \epsilon_k, \tag{2.24}$$

provided that it is held together by the strong or electromagnetic interaction. (See, however, §3.2.)

Another (universally conserved) multiplicative quantum number is the *statistics* defined for all particles so that is is $+1$ for bosons and -1 for fermions. According to the spin-statistics theorem (§1.14), statistics always equals $(-1)^{2s}$, where s is the spin. Thus, its conservation follows from the "modified" conservation law for the spin, i.e., the vector addition rule (2.23), where, necessarily, $\ell \in \mathbb{Z}$. For similar reasons, the strong interactions conserve the *isospin class*, which is the $\mathbb{Z}_2$ valued multiplicative quantum number defined to be $(-1)^{2I}$ for any hadron with isospin I.

For particle species that distinguish an orientation of space (or spacetime), namely, the neutrinos and antineutrinos, one calls the resulting orientation the *helicity* of the particle. Of course, the manner in which the orientation is assigned to the particle is conventional and can always be reversed. Fixing an orientation of space (e.g., with the aid of the right hand), one may regard helicities as numbers, usually $\frac{1}{2}$ or $-\frac{1}{2}$, where the factor of $\frac{1}{2}$ represents the spin of the neutrino. Then, ν_e, ν_μ, ν_τ have helicity $-\frac{1}{2}$, while the helicity of $\bar{\nu}_e, \bar{\nu}_\mu$ and $\bar{\nu}_\tau$ equals $\frac{1}{2}$.

Other important invariants are the $\mathrm{SU}(n)$ *spin* and $\mathrm{SU}(n)$ *class*, defined for all hadrons and valued in the semigroup $(\frac{1}{n}\mathbb{Z}_+)^{n-1}$ and the group $\mathbb{Z}_n$, respectively, n being any integer with $2 \leq n \leq 6$. For $n = 2$ they coincide with isospin and isospin class, while $\mathrm{SU}(3)$ spin is also referred to as *unitary spin*. See §§8.3, 8.6.

References: Ryder 1986; Sudbery 1986.

3 More on Relativistic Particle Models*

In this chapter we justify the choice of free matter-particle models described in §§1.16, 1.17. The argument is based on classifying the first-order natural bundles over spacetimes, i.e., representations of the isochronous Lorentz group, and then using analogy with the nonrelativistic case.

3.1 Reducible Representations of O(3)

Any irreducible, finite-dimensional, complex representation τ of the Lie algebra $\mathfrak{su}(2) \cong \mathfrak{so}(3)$ is classified, up to an equivalence, by its spin $s \in \{0, \frac{1}{2}, 1, \frac{3}{2}, \ldots\} = \frac{1}{2}\mathbb{Z}_+$ (§2.1). Moreover, by (2.4), for any $X \in \mathfrak{su}(2)$ which is unit with respect to a suitably normed Killing form, $i\tau(X)$ has the eigenvalues $-s, -s+1, \ldots, s-1, s$, all simple.

Whether or not τ is irreducible, it comes from a representation of the compact Lie group SU(2), and so the representation space admits a Hermitian inner product $\langle,\rangle$ which is τ-invariant, i.e., makes all endomorphisms in the image of τ skew-adjoint. In particular, every finite dimensional complex representation τ of $\mathfrak{su}(2) \cong \mathfrak{so}(3)$ is a direct sum of irreducible subrepresentations. The spins $s_1, \ldots, s_k$ (possibly with repetitions) of the irreducible summands of such a representation τ can be "intrinsically" characterized, in terms of τ alone, as follows. Arranging them so that $s_1 \geq \ldots \geq s_k$, fixing any unit $X \in \mathfrak{su}(2)$, and finding the eigenvalues of $i\tau(X)$ (which all lie in $\frac{1}{2}\mathbb{Z}$), along with their multiplicities, we see (§2.1) that s_1 is the highest eigenvalue, and it is realized in an irreducible subspace V of spin s_1. Removing from the spectrum of $i\tau(X)$ the eigenvalues $-s_1, -s_1+1, \ldots, s_1-1, s_1$ occurring each with multiplicity one in V, i.e., reducing their multiplicities by one, we obtain the spectrum of $i\tau(X)$ restricted to the τ-invariant subspace $V^\perp$ (orthogonal complement of V relative to $\langle,\rangle$ chosen as before), so that s_2 is the highest eigenvalue of the reduced spectrum, etc.

In other words, writing out the eigenvalues of $i\tau(X)$, each repeated as many times as its multiplicity indicates, and decomposing the resulting table into pairwise disjoint progressions of step 1, symmetric about 0, we find the $s_1, \ldots, s_k$ by selecting the largest number in each progression.

3.2 Tensor Products of Representations

Given Lie group representations $\varphi_\alpha : G_\alpha \to \operatorname{Aut} V_\alpha$, $\alpha = 1, \ldots, \ell$, in finite dimensional vector spaces V_α, we obtain a representation $\varphi : G_1 \times \ldots \times G_\ell \to \operatorname{Aut}(V_1 \otimes \ldots \otimes V_\ell)$ by setting $\varphi(h_1, \ldots, h_\ell) = \varphi(h_1) \otimes \ldots \otimes \varphi(h_\ell)$. The corresponding construction for Lie algebra representations $\tau_\alpha : \mathfrak{g}_\alpha \to \operatorname{End} V_\alpha$, $\alpha = 1, \ldots, \ell$, results in the representation $\tau : \mathfrak{g}_1 \oplus \ldots \oplus \mathfrak{g}_\ell \to \operatorname{End}(V_1 \otimes \ldots \otimes V_\ell)$ given by $\tau(a_1, \ldots, a_\ell) = \tau_1(a_1) \otimes \operatorname{Id}_{V_2} \otimes \ldots \otimes \operatorname{Id}_{V_\ell} + \operatorname{Id}_{V_1} \otimes \tau_2(a_2) \otimes \operatorname{Id}_{V_3} \otimes \ldots \otimes \operatorname{Id}_{V_\ell} + \ldots + \operatorname{Id}_{V_1} \otimes \ldots \otimes \operatorname{Id}_{V_{\ell-1}} \otimes \tau_\ell(a_\ell)$. In the case where $G_1 = \ldots = G_\ell = G$ or $\mathfrak{g}_1 = \ldots = \mathfrak{g}_\ell = \mathfrak{g}$, one defines the *tensor product* $\varphi_1 \otimes \ldots \otimes \varphi_\ell : G \to \operatorname{Aut}(V_1 \otimes \ldots \otimes V_\ell)$ of the representations $\varphi_\alpha : G \to \operatorname{Aut} V_\alpha$, as well as the tensor product $\tau_1 \otimes \ldots \otimes \tau_\ell : \mathfrak{g} \to \operatorname{End}(V_1 \otimes \ldots \otimes V_\ell)$ of the $\tau_\alpha : \mathfrak{g} \to V_\alpha$, to be the representation obtained by taking the *diagonal homomorphism* $G \to G \times \ldots \times G$ or $\mathfrak{g} \to \mathfrak{g} \oplus \ldots \oplus \mathfrak{g}$ followed by φ (or τ) defined above.

Now let τ_1, τ_2 be two irreducible finite dimensional complex representations of the Lie algebra $\mathfrak{su}(2) \cong \mathfrak{so}(3)$, with spins s_1, s_2 and representations spaces V_1, V_2. The irreducible summands of the tensor product representation $\tau_1 \otimes \tau_2 : \mathfrak{su}(2) \to \operatorname{End}(V_1 \otimes V_2)$ then have the spins

$$s_1 + s_2, \; s_1 + s_2 - 1, \ldots, \; |s_1 - s_2| + 1, \; |s_1 - s_2|, \tag{3.1}$$

each occurring only once. In fact, fixing any unit $X \in \mathfrak{su}(2)$ and choosing bases u_j of V_1, w_k of V_2, $j = -s_1, -s_1 + 1, \ldots s_1$, $k = -s_2, -s_2 + 1, \ldots, s_2$, with $i\tau_1(X)u_j = ju_j$, $i\tau_2(X)w_k = kw_k$ (cf. (2.4)), we obtain an eigenvector basis $u_j \otimes w_k$ for $i(\tau_1 \otimes \tau_2)(X)$ in $V_1 \otimes V_2$ with eigenvalues $j + k$ (where j, k vary as above). Our assertion is now immediate if we split the rectangular table (matrix) A with $A_{jk} = j + k$ into progressions as at the end of §3.1. For $s_1 = 2, s_2 = \frac{3}{2}$, this is illustrated by the following diagram, where dots stand for matrix entries, forming progressions as indicated by the connecting segments.

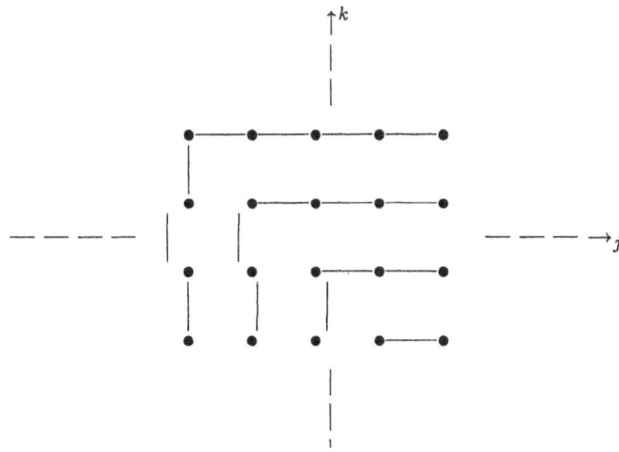

More generally, the set of spins of irreducible summands for $\tau_1 \otimes \ldots \otimes \tau_k$, where $\tau_1, \ldots, \tau_k$ are irreducible representations of $\mathfrak{su}(2)$ with spins $s_1, \ldots, s_k$, ordered so that $s_1 \geq \ldots \geq s_k$, consists of all numbers s with

$$s_1 + \ldots + s_k \geq s \geq \max(0, s_1 - s_2 - \ldots - s_k) \tag{3.2}$$

and

$$s_1 + \ldots + s_k - s \in \mathbb{Z}. \tag{3.3}$$

(Note that, if the order of $s_1, \ldots, s_k$ is arbitrary, $s_1 - s_2 - \ldots - s_k$ has to be replaced by $2 \max_{1 \leq j \leq k} s_j - \sum_{j=1}^{k} s_j$.) Irreducible summands with a given spin s may occur several times and we are not interested in their multiplicities. However, the most obvious argument we can use to prove the above assertion, namely induction on k along with the previously established case $k = 2$ (cf. (3.1)), also shows that the multiplicity of the *highest* spin $s_1 + \ldots + s_k$ equals one.

Instead of the Lie algebra $\mathfrak{su}(2) \cong \mathfrak{so}(3)$, we could of course regard the above assertion as a result on tensor products of irreducible representations of the group $SU(2)$ (or $SO(3)$, also allowing double-valued representations). The statements about spins also applies to irreducible, finite dimensional, complex representations of the two-component group $\overline{SU}(2)$ defined as in §2.1 (or $O(3)$, with possible double-valuedness). However, our discussion now also involves the parity ϵ of the representation, defined by the condition that the image of $\Pi \in \overline{SU}(2)$ (see §2.1) be $\epsilon \cdot \mathrm{Id}$ for bosonic representations (with spin $s \in \mathbb{Z}$) and $\epsilon i \cdot \mathrm{Id}$ for fermionic ones, with $s \notin \mathbb{Z}$ (cf. §2.2). Obviously, all irreducible summands for the tensor product of representations with parities $\epsilon_1, \ldots, \epsilon_k$ have the same parity ϵ, and either $\epsilon = \epsilon_1 \ldots \epsilon_k$ (if the number q of fermionic factors is congruent to 0 or 1 modulo 4), or $\epsilon = -\epsilon_1 \ldots \epsilon_k$ (in the case where $q \equiv 2$ or $q \equiv 3 \bmod 4$).

Describing the possible composite systems of particles with the aid of nonrelativistic models, using irreducible summands of tensor products of natural bundles over a Riemannian 3-manifold $(\mathbb{M}, g)$, and also including orbital angular momentum (in full analogy with the *relativistic* discussion in §§1.10, 1.13), we thus obtain from (3.2) and (3.3) the vector addition rule (2.23) for the spins and isospins. On the other hand the "experimental multiplication rule" (2.24) for parities is realized by such models *only* if the number of fermions involved equals 0 or 1 modulo 4. This discrepancy can be removed by modifying fermionic representations as in Example 2.3.6, and defining the parity ϵ by requiring that the image of $\Pi \in \widetilde{SU}(2)$ be $\epsilon \cdot \mathrm{Id}$. However, that definition in turn contradicts the experimental fact that antiparticle formation (which replaces the representation space by its complex conjugate) *changes* the parity of fermions (§1.14).

Consequently, in contrast with spin, the formal description of nonrelativistic parity is plagued with problems. Some of them may be removed by using a relativistic approach, but they are then actually traded for difficulties of another type (see §3.9).

References: Zhelobenko 1973; Mackey 1963; Schweber 1962.

3.3 The Lorentz Group

Let V be a *Lorentz vector space*, that is, a 4-dimensional real vector space endowed with a fixed *Lorentz inner product* $\langle , \rangle$ (a nondegenerate bilinear form of signature $-+++$). Vectors $v \in V$ with $\langle v, v \rangle < 0$ (or $\langle v, v \rangle = 0$, or $\langle v, v \rangle > 0$) then are referred to as *timelike* (or *null*, or *spacelike*). The set of timelike vectors in V has two connected components, known as the *time orientations* of V. (From now on we will simply write V instead of $(V, \langle , \rangle)$.) The *linear isometries* of V onto itself, i.e., automorphisms that leave $\langle , \rangle$ invariant, form a Lie group denoted $O(V)$, each of whose elements either preserves both time orientations, or interchanges them. The *Lorentz group* $O^\uparrow(V)$ of V is the open index 2 subgroup of $O(V)$, consisting of all time-orientation preserving transformations. (Sometimes $O^\uparrow(V)$ is called the *isochronous* Lorentz group of V.) The group $O^\uparrow(V)$ contains the index 2 subgroup $SO^\uparrow(V)$ formed by its orientation-preserving elements. We also use the notations $O(3,1)$, $O^\uparrow(3,1)$, $SO^\uparrow(3,1)$ for the analogous *matrix* groups, obtained when V is $\mathbb{R}^4$ with the standard Lorentz inner product.

We usually assume that V is *time-oriented*, i.e., there has been made a choice of one time orientation in V (cf. §1.1), the elements of which then are called the *future-pointing* (timelike) vectors. For any fixed future-pointing unit timelike vector $B_0 \in V$, the orthogonal complement $B_0^\perp$ of B_0 is a Euclidean 3-space with the inner product induced from $\langle , \rangle$, and the isotropy group $O(B_0^\perp) \cong O(3)$ of B_0 in $O^\uparrow(V)$ is a deformation retract of $O^\uparrow(V)$, so that $O^\uparrow(V)$ has the homotopy type of $O(3)$. This follows from the fact that $O(B_0^\perp)$ is a maximal compact subgroup of $O^\uparrow(V)$, but can also be obtained from the deformation of Lorentz-orthonormal bases of V with future-pointing first vectors B, onto such bases with the first vector B_0, given by the obvious hyperbolic rotation ("boost") of B towards B_0. Therefore, $\pi_1 SO(3,1) \cong \mathbb{Z}_2$ (cf. §2.1).

References: Ramond 1981; Schweber 1962.

3.4 Infinitesimal Representations of $O(3,1)$

Let $\mathfrak{so}(V)$ be the Lie algebra of $SO^\uparrow(V)$ (or $O^\uparrow(V)$, $O(V)$) for a Lorentz vector space V (§3.3), with its matrix-algebra counterpart $\mathfrak{so}(3,1)$. To study representations of $O^\uparrow(V)$, we first look at them infinitesimally, i.e., discuss (finite dimensional) representations of $\mathfrak{so}(V)$. Since it is *complex* linear representations that we are interested in, we may as well concentrate on representations of the complex Lie algebra $\mathfrak{so}(V) \otimes \mathbb{C}$, obtained by complexifying $\mathfrak{so}(V)$. In general, representations $\tau : \mathfrak{g} \to \operatorname{End} W$ of any real Lie algebra $\mathfrak{g}$ in a complex vector space W are in a bijective correspondence with representations $\tilde\tau : \mathfrak{g} \otimes \mathbb{C} \to \operatorname{End} W$ of its complexification $\mathfrak{g} \otimes \mathbb{C}$. The correspondence arises by defining $\tilde\tau$ to be the unique complex-linear extension of τ, as a map between Lie algebras. Also, when τ is regarded as a representation of $\mathfrak{g}$ in the *conjugate* vector space $\overline{W}$ of W (§1.6), i.e., we consider the *conjugate representation* $\overline\tau$ of τ, then the corresponding extension $\mathfrak{g} \otimes \mathbb{C} \to \operatorname{End} W = \operatorname{End} \overline{W}$ equals $\tilde\tau \circ (\ )^-$, $(\ )^-$ being the complex conjugation in $\mathfrak{g} \otimes \mathbb{C}$. Thus,

$$(\overline\tau)^{\widetilde{}} = \tilde\tau \circ (\ )^-. \tag{3.4}$$

The complexification $\mathfrak{so}(V) \otimes \mathbb{C}$ of $\mathfrak{so}(V)$ can in turn be naturally identified with $\mathfrak{o}(V \otimes \mathbb{C})$, the complex Lie algebra of all complex linear endomorphisms of the complexification $V \otimes \mathbb{C}$ of V that are skew-adjoint with respect to the unique complex-bilinear extension of $\langle,\rangle^{\mathbb{C}}$ of $\langle,\rangle$ to $V \otimes \mathbb{C}$. (Thus, $\mathfrak{o}(V \otimes \mathbb{C})$ is the Lie algebra of the group $O(V \otimes \mathbb{C})$ of all complex linear automorphisms of $V \otimes \mathbb{C}$ that preserve $\langle,\rangle^{\mathbb{C}}$.) In fact, the inclusion $\mathfrak{so}(V) \subset \mathfrak{o}(V \otimes \mathbb{C})$ defined by the unique complex linear extension of automorphisms of V gives rise to $\mathfrak{so}(V) \otimes \mathbb{C} \subset \mathfrak{o}(V \otimes \mathbb{C})$, and so the latter two algebras have to coincide as they have equal dimensions.

Consequently, $\mathfrak{so}(V) \otimes \mathbb{C} = \mathfrak{o}(V \otimes \mathbb{C}) = \mathfrak{so}(V') \otimes \mathbb{C}$ for any other *real form* $(V', \langle,\rangle')$ of $(V \otimes \mathbb{C}, \langle,\rangle^{\mathbb{C}})$, that is, any real subspace V' of $V \otimes \mathbb{C}$ such that the restriction $\langle,\rangle'$ of $\langle,\rangle^{\mathbb{C}}$ to V' is real-valued and nondegenerate. Choosing V' which is *Euclidean* in the sense that $\langle,\rangle$ is, in addition, positive definite, we see that $\mathfrak{so}(3,1)$ and $\mathfrak{so}(4)$ have isomorphic complexifications.

The structures of $SO(4)$ and $SO(3)$, as well as their Lie algebras, can easily be described in terms of the normed algebra $\mathbb{H}$ of quaternions and the multiplicative unit quaternion group $S^3 \cong SU(2)$ (see §2.1). In fact, the orthogonal decomposition $\mathbb{H} = \operatorname{Re}\mathbb{H} \oplus \operatorname{Pu}\mathbb{H}$ with $\operatorname{Re}\mathbb{H} = \mathbb{R}1$, $\operatorname{Pu}\mathbb{H} = 1^\perp$, where $1 \in \mathbb{H}$ is the unit, leads to the group homomorphism

$$\Phi_3 : S^3 \to SO(\operatorname{Pu}\mathbb{H}) \cong SO(3)$$

with $[\Phi_3(q)]u = quq^{-1}$ (quaternion multiplication) for $q \in S^3, u \in \operatorname{Pu}\mathbb{H}$. Similarly, we have the homomorphism

$$\Phi_4 : S^3 \times S^3 \to SO(\mathbb{H}) \cong SO(4)$$

with $[\varPhi_4(q_1, q_2)]u = q_1 u q_2^{-1}$. Both $\varPhi_3, \varPhi_4$ are two-fold coverings, as they have two-element kernels and the groups they act between are of equal dimensions (cf. §2.1). Thus, they induce Lie algebra isomorphisms

$$\mathfrak{su}(2) \cong \mathfrak{s}^3 \cong \mathfrak{so}(3)$$

and

$$\mathfrak{su}(2) \oplus \mathfrak{su}(2) \cong \mathfrak{s}^3 \oplus \mathfrak{s}^3 \cong \mathfrak{so}(4)$$

as well as the commuting diagram

$$
\begin{array}{ccc}
S^3 & \xrightarrow{\varPhi_3} & SO(3) \\
{\scriptstyle \text{diagonal}} \downarrow & & \downarrow {\scriptstyle \begin{array}{c}\text{trivial}\\\text{extension}\end{array}} \\
S^3 \times S^3 & \xrightarrow{\varPhi_4} & SO(4).
\end{array}
\tag{3.5}
$$

Complex linear representations of $\mathfrak{so}(3, 1)$ are thus in a bijective correspondence with those of $\mathfrak{so}(4)$, since both Lie algebras have the same complexification (see above). For such representations that are finite dimensional, every invariant subspace has an invariant complement, as $\mathfrak{so}(4)$ is the Lie algebra of the compact simply connected Lie group $S^3 \times S^3$. Therefore, each such representation of $\mathfrak{so}(V) \cong \mathfrak{so}(3, 1)$ is a direct sum of irreducible subrepresentations. (This, of course, remains true for $\mathfrak{so}(3, 1)$ and $S^3 \times S^3$ replaced by $\mathfrak{so}(k, \ell)$ and Spin (n), whenever $n = k + \ell \geq 3$.)

Finite dimensional, irreducible complex representations of $\mathfrak{so}(V) \cong \mathfrak{so}(3, 1)$, being exactly those of $\mathfrak{so}(4) \cong \mathfrak{su}(2) \oplus \mathfrak{su}(2)$, now can be classified as follows. Such a representation τ has to be obtained from two irreducible representations of the $\mathfrak{su}(2)$ summands by the tensor-product construction of §3.2. In fact, choose any unit $X_1, X_2 \in \mathfrak{su}(2)$ and find the lowest eigenvalue $-s_2$ of $i\tau(0, X_2)$ realized in the $0 \oplus \mathfrak{su}(2)$-invariant eigenspace of $i\tau(X_1, 0)$ for its lowest eigenvalue $-s_1$. (Thus, $s_1, s_2 \in \frac{1}{2}\mathbb{Z}_+$, cf. §2.1.) Fix $x_{s_1, s_2} \neq 0$ in the representation space with

$$i\tau(X_1, 0)x_{s_1, s_2} = -s_1 x_{s_1, s_2}, \quad i\tau(0, X_2)x_{s_1, s_2} = -s_2 x_{s_1, s_2}$$

and let $\varPhi_\alpha^\pm = Y_\alpha \pm iZ_\alpha \in \mathfrak{su}(2) \otimes \mathbb{C}$ (cf. §2.1), Y_1, Y_2 and Z_1, Z_2 being chosen so that the bases $X_\alpha, Y_\alpha, Z_\alpha$ of $\mathfrak{su}(2)$, $\alpha = 1, 2$, are both positive oriented and orthonormal. For $j = -s_1, -s_1 + 1, \ldots, s_1$ and $k = -s_2, -s_2 + 1, \ldots, s_2$, the formula

$$x_{j,k} = \tau(\varPhi_1^+, 0)^{j+s_1} \tau(0, \varPhi_2^+)^{k+s_2} x_{s_1, s_2}$$

defines a basis of a τ-invariant space with

$$i\tau(X_1, 0)x_{j,k} = -j x_{j,k}, \quad i\tau(0, X_2)x_{j,k} = -k x_{j,k},$$

thus proving our assertion, with the factor spaces spanned by $x_{s_1,s_2}, x_{s_1-1,s_2},$ $\ldots, x_{-s_1,s_2}$ and $x_{s_1,s_2}, x_{s_1,s_2-1}, \ldots, x_{s_1,-s_2}$, respectively.

Irreducible finite-dimensional complex representations τ of $\mathfrak{so}(V) \cong \mathfrak{so}(3,1)$ thus are classified by the pairs of spins (§2.1)

$$(s_1, s_2) \in \frac{1}{2}\mathbb{Z}_+ \times \frac{1}{2}\mathbb{Z}_+.$$

Furthermore, if $\mathfrak{so}(B_0^\perp) \subset \mathfrak{so}(V)$ is the isotropy algebra of a unit time-like $B_0 \in V$, then the representation τ with s_1, s_2 as above, restricted to $\mathfrak{so}(B_0^\perp) \cong \mathfrak{su}(2)$, is a direct sum of irreducible representations with spins

$$s_1 + s_2, \quad s_1 + s_2 - 1, \ldots, \quad |s_1 - s_2|$$

each occurring just once. In fact, $\mathfrak{so}(B_0^\perp) \otimes \mathbb{C} \subset \mathfrak{so}(V) \otimes \mathbb{C}$ is the isotropy algebra of $B_0 \in V \subset V \otimes \mathbb{C}$ in $\mathfrak{so}(V \otimes \mathbb{C})$, due to the obvious inclusion and equal dimensionalities, and hence it is also the complexification of the isotropy algebra for a Euclidean space form V', chosen so as to contain B_0. The latter algebra may be identified with the diagonal of $\mathfrak{su}(2) \oplus \mathfrak{su}(2)$ (by (3.5)), and our assertion follows from (3.1).

Let us also note that the invariants s_1, s_2 classifying finite dimensional, irreducible complex representations of $\mathfrak{so}(3,1)$, i.e., of $\mathfrak{so}(3,1) \otimes \mathbb{C} \cong (\mathfrak{su}(2) \otimes \mathbb{C}) \oplus (\mathfrak{su}(2) \otimes \mathbb{C})$, can be characterized in terms of the direct summands isomorphic to $\mathfrak{su}(2) \otimes \mathbb{C}$ (which in turn are easily seen to be the only two nonzero proper ideals of $\mathfrak{so}(3,1) \otimes \mathbb{C}$). Namely, the representation, restricted to one of them, becomes a direct sum of $2s_2 + 1$ copies of the irreducible, spin s_1 representation of $\mathfrak{su}(2)$, while, restricted to the other, it splits into $2s_1 + 1$ copies of the spin s_2 representation. (Recall that representations of $\mathfrak{su}(2) \otimes \mathbb{C}$ are the same as complex representations of $\mathfrak{su}(2)$.) In particular, the representation space corresponding to (s_1, s_2) is of dimension $(2s_1 + 1)(2s_2 + 1)$. It is also clear that the distinction between s_1 and s_2 is conventional, as it is based solely on the *order* of the direct summands. Actually, s_1 and s_2 become interchanged when the representation is replaced by its complex conjugate, cf. §3.6.

The number $s_1 + s_2$ is called the *spin* of the irreducible representation τ of $\mathfrak{so}(3,1)$ characterized by the invariants (s_1, s_2), which is clearly justified by the fact that it is the largest ("dominant") spin among those of the irreducible summands of τ restricted to the isotropy algebra $\mathfrak{so}(3)$ of any timelike vector.

We say that an irreducible representation of $\mathfrak{so}(3,1)$ is *bosonic (fermionic)* if its spin $s_1 + s_2$ is an integer (a non-integral half-integer). Thus, a representation is bosonic or fermionic if and only if so are the irreducible summands of its restriction to the isotropy algebra $\mathfrak{so}(3)$ of any timelike vector (see §2.1).

References: Ramond 1981; Schweber 1962; Hayward 1976.

3.5 The Role of SL(2, ℂ)

Let us now consider a Weyl pair (S, Ω) along with its associated Lorentz vector space T (see §1.3). The group $\mathrm{SL}(S) \cong \mathrm{SL}(2, \mathbb{C})$ of all complex linear automorphisms of S that leave Ω invariant, i.e., have determinant 1, has a natural $\mathbb{Z}_2$ extension $\overline{\mathrm{SL}}(S)$, for which $\overline{\mathrm{SL}}(S) \smallsetminus \mathrm{SL}(S)$ consists of all anti-automorphisms of S sending Ω onto $\overline{\Omega}$. Any $F \in \overline{\mathrm{SL}}(S)$ acts on T as the time-orientation preserving linear (Lorentz) isometry

$$T \ni B \mapsto FBF^{-1} \in T,$$

which is orientation-preserving if and only if $F \in \mathrm{SL}(S)$. The resulting homomorphisms $\overline{\mathrm{SL}}(S) \to \mathrm{O}^\uparrow(T)$, $\mathrm{SL}(S) \to \mathrm{SO}^\uparrow(T)$ are twofold coverings (for the same reasons as in §3.4). In particular, $\mathrm{SL}(S) \cong \mathrm{SL}(2, \mathbb{C})$ is simply connected, as $\pi_1 \mathrm{SO}(3,1) \cong \mathbb{Z}_2$ (§3.3).

Representations (possibly multi-valued) of $\mathrm{O}^\uparrow(T) \cong \mathrm{O}^\uparrow(3,1)$ are, therefore, at most double-valued and can always be regarded as single-valued representations of $\overline{\mathrm{SL}}(S)$ (of course, provided that one removes their additional, "artificial" branches). Finite dimensional complex representations (in the same sense) of the connected groups $\mathrm{SL}(S)$, $\mathrm{SO}^\uparrow(T)$ are classified by representations of the Lie algebra $\mathfrak{sl}(S) \cong \mathfrak{so}(T)$, and so each of them is a direct sum of irreducible subrepresentations, characterized in turn by their spin invariants (s_1, s_2) (§3.4). Consequently, a similar decomposition into irreducible summands is always possible for representations of $\overline{\mathrm{SL}}(S)$ or $\mathrm{O}^\uparrow(T)$.

Although $\mathrm{SL}(2, \mathbb{C})$ is a complex Lie group (of dimension 3), we regard it as real and the representations $\mathrm{SL}(2, \mathbb{C}) \to \mathrm{Aut}\,_\mathbb{C} W$ that we study are assumed to be *real* analytic homomorphisms (that happen to act between complex Lie groups). Analogously, on the Lie algebra level, the representations $\mathfrak{sl}(2, \mathbb{C}) \to \mathrm{End}\,_\mathbb{C} W$ are not required to be complex-linear.

An irreducible representation of $\mathrm{SL}(S) \cong \mathrm{SL}(2, \mathbb{C})$ descends to a (single-valued) representation of $\mathrm{SO}^\uparrow(T) \cong \mathrm{SO}^\uparrow(3,1)$ if and only if its spin is an integer. This is immediate from the fact that a representation τ of $\mathfrak{so}(3,1)$ which is irreducible and has spin invariants (s_1, s_2), restricted to the isotropy algebra $\mathfrak{so}(3)$ of a timelike vector has only irreducible summands with spins congruent to $s_1 + s_2$ modulo $\mathbb{Z}$ (§3.4), while $\mathfrak{so}(3)$ corresponds to the isotropy group $\mathrm{SO}(3)$.

References: Ramond 1981; Schweber 1962.

3.6 Conjugate Representations

Consider a representation φ of $\mathrm{SL}(S) \cong \mathrm{SL}(2, \mathbb{C})$ as in §3.5 (or, a possibly double-valued representation φ of $\mathrm{SO}^\uparrow(3,1) \cong \mathrm{SO}^\uparrow(T)$) in a finite dimensional complex vector space W. If φ is irreducible and corresponds to the spin invariants (s_1, s_2), then either of the following two representations:

(i) The conjugate $\overline{\varphi} : \mathrm{SL}(S) \to \mathrm{Aut}\,\overline{W} = \mathrm{Aut}\,W$ (the latter equality being between *real* Lie groups, cf. §3.5),

(ii) The composite $\varphi \circ \mathrm{ad}_{F_0} : \mathrm{SL}(S) \to \mathrm{Aut}\,W$ of φ with the inner automorphism $\mathrm{ad}_{F_0} \in \mathrm{Aut}\,(\mathrm{SL}(S))$ for any $F_0 \in \overline{\mathrm{SL}}(S) \smallsetminus \mathrm{SL}(S)$,

is (obviously) irreducible, and has the *reversed* spin invariants (s_2, s_1). Clearly, $\mathrm{SL}(S)$ can be replaced in the above discussion by $\mathrm{SO}^\uparrow(T)$.

To prove this, let us embed the quaternion normed algebra $\mathbb{H}$ into $T \otimes \mathbb{C}$ as a Euclidean real form, by choosing a unit timelike $B_0 \in T$ along with linear isometries $\mathrm{Pu}\,\mathbb{H} \to B_0^\perp \subset T \subset T \otimes \mathbb{C}$, $\mathrm{Re}\,\mathbb{H} \to i\mathbb{R}B_0 \subset iT \subset T \otimes \mathbb{C}$ (notation of §3.4). The Lie algebra representation $\overline{\tau}$ corresponding to $\overline{\varphi}$ has the property that its complex linear extension $\widetilde{\overline{\tau}}$ to $\mathfrak{so}(T) \otimes \mathbb{C}$ equals $\widetilde{\tau} \circ (\ )^-$ (see (3.4)). The conjugation $(\ )^-$ applied to any endomorphism a of $T \otimes \mathbb{C}$ (including elements of $\mathfrak{so}(T) \otimes \mathbb{C} = \mathfrak{o}(T \otimes \mathbb{C})$) results in $\overline{a} = (\ )^- \circ a \circ (\ )^-$, where $(\ )^-$ in the last expression acts on $T \otimes \mathbb{C}$ and is characterized as the unique antilinear extension of Id_T. Obviously, $(\ )^-$ leaves $\mathbb{H}$ embedded into $T \otimes \mathbb{C}$ invariant and, restricted to $\mathbb{H}$, equals minus the quaternion conjugation $(\ )^*$. The assertion for (i) now follows as $(\ )^*$ is antimultiplicative, while the summands in $\mathfrak{so}(T) \otimes \mathbb{C} \cong [\mathfrak{su}(2) \otimes \mathbb{C}] \oplus [\mathfrak{su}(2) \otimes \mathbb{C}]$ correspond to actions of S^3 on $\mathbb{H}$ by left and right multiplications, respectively, extended to $T \otimes \mathbb{C} = \mathbb{H} \oplus i\mathbb{H}$ by $\mathbb{C}$-linearity.

To establish the assertion for (ii), note that for $F_0 \in \mathrm{SL}(S)$, $\varphi \circ \mathrm{ad}_{F_0}$ is equivalent to φ, as a representation of $\mathrm{SL}(S)$ (the equivalence being $\varphi(F_0)$), so that $\varphi \circ \mathrm{ad}_{F_0}$ for $F_0 \in \overline{\mathrm{SL}}(S) \smallsetminus \mathrm{SL}(S)$ are all mutually equivalent. Therefore, we may choose F_0 to be a preimage in $\overline{\mathrm{SL}}(S)$ of the reflection in $\mathbb{R}B_0$, equal to Id on $\mathbb{R}B_0$ and to $-\mathrm{Id}$ on $B_0^\perp$ (as B_0 is unit, the preimages in question are actually B_0 and $-B_0$, which as elements of $T \subset \mathrm{End}_\mathbb{R}S$ happen to lie in $\overline{\mathrm{SL}}(S)$). We may obviously replace φ by the corresponding Lie algebra representation $\tau : \mathfrak{so}(T) \to \mathrm{End}\,W$, or its $\mathbb{C}$-linear extension $\widetilde{\tau} : \mathfrak{so}(T \otimes \mathbb{C}) \to \mathrm{End}\,W$. For $a \in \mathfrak{so}(T \otimes \mathbb{C})$, $\mathrm{ad}_{F_0}(a) = R_0 \circ a \circ R_0$, R_0 being the reflection in $\mathbb{R}B_0$, extended $\mathbb{C}$-linearly to $T \otimes \mathbb{C}$. Note that R_0 leaves $\mathbb{H}$ invariant and restricted to $\mathbb{H}$ equals the quaternion conjugation. Our assertion for (ii) now follows exactly as for (i).

References: Ramond 1981; Schweber 1962.

3.7 Representations of $O^\uparrow(3,1)$

We now proceed to classify irreducible, finite-dimensional, complex representation of $\overline{SL}(2,\mathbb{C}) \cong \overline{SL}(S)$ (or, analogous representations of $O^\uparrow(3,1) \cong O^\uparrow(T)$, including double-valued ones), using what we know about representations of $SL(2,\mathbb{C})$ (§§3.5, 3.4). For simplicity, set $G = \overline{SL}(2,\mathbb{C})$, $G^0 = SL(2,\mathbb{C})$. In order that a given representation $\varphi_0 : G^0 \to \operatorname{Aut} W$ admit an extension to a representation $\varphi : G \to \operatorname{Aut} W$, it is necessary and sufficient that there exist $A \in \operatorname{Aut} W$ such that, for some $F_0 \in G \setminus G_0$,

(a) A is an equivalence between the representations φ_0 and $\varphi_0 \circ \operatorname{ad}_{F_0}$ of G^0 in W,
(b) $A^2 = \varphi_0(F_0^2)$.

(This assertion remains true for any group G with an index 2 subgroup G^0.) In fact, necessity, with $A = \varphi(F_0)$ is obvious (for *each* $F_0 \in G \setminus G^0$), while sufficiency is obtained by requiring that $A = \varphi(F_0)$, i.e., setting $\varphi(F) = A\varphi_0(F_0^{-1}F)$ for all $F \in G \setminus G^0$.

Examples of irreducible representations $\varphi : G \to \operatorname{Aut} W$ of $G = \overline{SL}(2,\mathbb{C})$ now can be constructed as follows.

(i) Given $s_1 \in \frac{1}{2}\mathbb{Z}_+$, let $\varphi_0 : G^0 \to \operatorname{Aut} W$ be an irreducible representation of $G^0 = SL(2,\mathbb{C})$ with spin invariants (s_1, s_1). Fix $F_0 \in G \setminus G^0$. Since φ_0 and $\varphi_0 \circ \operatorname{ad}_{F_0}$ then are equivalent (cf. (ii) of §3.6), we may choose an equivalence B between them. Consequently, B^2 is an equivalence between φ_0 and $\varphi_0 \circ \operatorname{ad}_{F_0^2}$ and, since so is $\varphi_0(F_0^2)$, we have $B^2 = z\varphi_0(F_0^2)$ with some $z \in \mathbb{C}$. (Here and in the sequel we repeatedly use the fact that an equivalence between two irreducible representations is unique up to a factor, which is an obvious consequence of Schur's lemma.) Setting $A = \frac{1}{u}B$, where $u^2 = z$, we obtain conditions (a), (b) and it is clear that these conditions, for a fixed F_0, determine A *uniquely up to a sign*. Thus, we may extend φ_0 in exactly two ways to representations φ of G in W, differing by the sign of A. These representations are not equivalent, as an equivalence between them would be a self-equivalence of φ_0 and hence a multiple of Id, so that it could not anticommute with A.

(ii) Let $s_1, s_2 \in \frac{1}{2}\mathbb{Z}_+$ be different and choose irreducible representations φ_0^+, φ_0^- of $G^0 = SL(2,\mathbb{C})$ in complex vector spaces W_+, W_- with spin invariants (s_1, s_2) and (s_2, s_1), respectively. (By (i) of §3.6, we may take, e.g., $W_- = \overline{W}_+$, $\varphi_0^- = \overline{\varphi}_0^+$.) Fix $F_0 \in G \setminus G_0$. By (ii) of §3.6, we may find equivalences $B_\pm : W_\pm \to W_\mp$ between $\varphi_0^\pm$ and $\varphi_0^\mp \circ \operatorname{ad}_{F_0}$. Then, $B_\mp B_\pm = z^\pm \varphi_0^\pm(F_0^2)$ as both $B_\mp B_\pm$ and $\varphi_0^\pm(F_0^2)$ are equivalences between the same pair of representations, $\varphi_0^\pm$ and $\varphi_0^\pm \circ \operatorname{ad}_{F_0^2}$. The numbers $z^\pm$ have to coincide, $z^- = z^+$, as $z^- \varphi_0^-(F_0^2)B_+ = (B_+B_-)B_+ = B_+(B_-B_+) = B_+ \cdot z^+ \varphi_0^+(F_0^2) = z^+ \varphi_0^-(F_0F_0^2F_0^{-1})B_+ = z^+\varphi_0^-(F_0^2)B_+$. Setting, e.g., $A_+ = B_+, A_- = \frac{1}{z^+}B_-$ and defining $A \in \operatorname{Aut}(W_+ \oplus W_-)$

by $A = A_{\pm}$ on $W_{\pm}$, we therefore obtain (a), (b) for $\varphi_0 = \varphi_0^+ \oplus \varphi_0^-$ and $W = W_+ \oplus W_-$, which results in an extension of φ_0 to a representation φ of $G = \overline{\mathrm{SL}}(2, \mathbb{C})$ in $W = W_+ \oplus W_-$. Note that our choice of $A = A_+ \oplus A_-$ is unique up to replacing it by $\tilde{A} = (uA_+) \oplus (\frac{1}{u}A_-)$ with $u \in \mathbb{C} \smallsetminus \{0\}$. However, using $\tilde{A}$ instead of A will only result in a representation $\tilde{\varphi}$ *equivalent* to φ, an equivalence being $\mathrm{Id}_{W_+} \oplus u \cdot \mathrm{Id}_{W_-}$. Irreducibility of φ is immediate as the only nonzero proper invariant subspaces of its restriction φ_0 to G^0 are the summands $W_{\pm}$ (which follows from a projection argument, φ_0^+ and φ_0^- being nonequivalent), while $W_{\pm}$ themselves are not invariant under $A = \varphi(F_0)$.

We have thus described, for each *unordered* pair $\{s_1, s_2\}$ with $s_1, s_2 \in \frac{1}{2}\mathbb{Z}_+$, a single irreducible representation of $\overline{\mathrm{SL}}(2, \mathbb{C})$ (if $s_1 \neq s_2$), or two such nonequivalent representations, if $s_1 = s_2$. In either case, $\{s_1, s_2\}$ will be referred to as the *spin invariants* of each resulting representation, while $s_1 + s_2$ will be called its *spin*. Again, we say that the representations is *bosonic* if $s_1 + s_2 \in \mathbb{Z}$, and *fermionic* otherwise. Bosonic representations are precisely those that descend to $\mathrm{O}^{\uparrow}(3, 1) = \overline{\mathrm{SL}}(2, \mathbb{C})/\mathbb{Z}_2$, i.e., form (single-valued) representations of $\mathrm{O}^{\uparrow}(3, 1)$ (§§3.4, 3.5). Those bosonic representations φ with $s_1 = s_2$ have an additional invariant, the *parity* $\epsilon \in \{1, -1\}$, defined to be the parity of the direct summand with highest spin (equal to $s_1 + s_2$) of the restriction of φ to the isotropy group $\mathrm{O}(3)$ of some (any) timelike vector B_0. Thus, ϵ does not depend on B_0 for reasons of continuity. Parity obviously changes when φ is tensor-multiplied by the representation $\mathrm{O}^{\uparrow}(3, 1) \rightarrow \mathrm{O}^{\uparrow}(3, 1)/\mathrm{SO}^{\uparrow}(3, 1) = \mathbb{Z}_2 = \{\mathrm{Id}, -\mathrm{Id}\}$ of $\mathrm{O}^{\uparrow}(3, 1)$ in the line of *pseudoscalars*, which for our purposes can be identified with exterior 4-forms on the underlying Lorentz vector space.

We will now show that every irreducible representation $\varphi : G \rightarrow \mathrm{Aut}\, W$ of $G = \overline{\mathrm{SL}}(2, \mathbb{C})$ is equivalent to one of those defined in (i), (ii) above, so that such representations are classified by the unordered pair of spin invariants $\{s_1, s_2\}$ and, if $s_1 = s_2$, also by the parity ϵ. Given such φ and a fixed $F_0 \in G \smallsetminus G^0$, let $W_+ \subset W$ be an irreducible invariant subspace of the restriction φ_0 of φ to G^0. Obviously, $W_- = AW_+$ with $A = \varphi(F_0)$ is also φ_0-invariant and irreducible, and so the φ-invariant span of W_+ and W_- coincides with W. By φ_0-irreducibility, W_+ and W_- either intersect trivially, or coincide, so that two cases are possible:

(I) $W_+ \cap W_- = \{0\}$, i.e., $W = W_+ \oplus W_-$. Obviously, φ is obtained as in (ii) above with $A = \varphi(F_0)$, $\varphi_0^{\pm}$ being the representations φ_0 defines in $W_{\pm}$, except that we have yet to prove that $s_1 \neq s_2$, for the spin invariants (s_1, s_2) of φ_0^+. (This will also show that in the case where $s_1 = s_2$, the construction of (ii) leads to a *reducible* representation.) In fact, suppose that $s_1 = s_2$. By (i), either of $\varphi_0^{\pm}$ can be extended to a representation $\varphi^{\pm}$ of G in $W_{\pm}$, by finding $C_{\pm} = \varphi^{\pm}(F_0)$ with $C_{\pm}^2 = \varphi^{\pm}(F_0^2)$ that establish equivalences between $\varphi_0^{\pm}$ and $\varphi_0^{\pm} \circ \mathrm{ad}_{F_0}$ in $W_{\pm}$, which makes either of $C_{\pm}$ unique up to a sign. Since $AC_- A^{-1}$

satisfies the conditions characterizing C_+, we have $AC_-A^{-1} = \epsilon_+C_+$ and, similarly, $AC_+A^{-1} = \epsilon_-C_-$, with $\epsilon_+, \epsilon_- \in \{1, -1\}$. Hence $A^2C_+ = \epsilon_-AC_-A = \epsilon_+\epsilon_-C_+A^2$ and, applying the relations $A^2 = \varphi_0^+(F_0^2)$, $C_+ = \varphi_0^+(F_0)$, we find that $\epsilon_+ = \epsilon_-$. Thus, A commutes with $C = (\epsilon_+C_+) \oplus C_-$ and hence, as both C and A are equivalences between φ_0 and $\varphi_0 \circ \mathrm{ad}_{F_0}$, $C^{-1}A$ commutes with φ (since it does with φ_0, and with $A = \varphi(F_0)$). Therefore, A is a multiple of C and so it leaves W_+ and W_- invariant, contradicting the relation $W_- = AW_+$. We have thus shown that $s_1 \neq s_2$ whenever $W_+ \cap W_- = \{0\}$ and φ is irreducible.

(II) In the remaining case, $W_+ = W_- = W$, so that $s_1 = s_2$, as $A = \varphi(F_0)$ establishes an equivalence between φ_0 and $\varphi_0 \circ \mathrm{ad}_{F_0}$ (cf. (ii) of §3.6). Hence φ is one of just two possible extensions of φ_0 due to the uniqueness assertion of (i) in this section.

Instead of $\overline{\mathrm{SL}}(2, \mathbb{C})$ we could also have used another twofold covering group of $\mathrm{O}^\uparrow(3, 1)$, obtained from $\overline{\mathrm{SL}}(2, \mathbb{C})$ as in Example 2.3.6 (we can denote it $\widetilde{\mathrm{SL}}(2, \mathbb{C})$). Again, complex representations of both groups are related by a natural bijective correspondence.

References: Ramond 1981; Schweber 1962; Hayward 1976.

3.8 Some Geometric Examples

Elementary algebraic arguments lead to the following constructions.

Example 3.8.1. For any Lorentz vector space V, the obvious representation of $\mathrm{O}^\uparrow(V)$ in the complexification $S_0^k V^* \otimes \mathbb{C}$ of the space $S_0^k V^*$ of all k-linear symmetric forms of contraction zero (i.e., degree k pseudo-spherical harmonics) on V, is irreducible, with spin invariants $\{\frac{k}{2}, \frac{k}{2}\}$ and parity $\epsilon = (-1)^k$. The corresponding representation in $S_0^k V^* \otimes \Lambda^4 V^* \otimes \mathbb{C}$ only differs from the previous one by parity (see §3.7).

Example 3.8.2. For any Majorana triple $(S, \pm I, \omega)$ and T as in §1.3, the natural representation of $\overline{\mathrm{SL}}(S)$ in the subspace of $S_0^k T^* \otimes S \otimes \mathbb{C}$ obtained by requiring that the Clifford multiplication (1.5) involving S and one of the T^* arguments be zero, is irreducible, with spin invariants $\{\frac{k+1}{2}, \frac{k}{2}\}$.

Example 3.8.3. Let Θ be the volume form of a Lorentz vector space V with a fixed orientation, so that $\Theta \in \Lambda^4 V^*$ is characterized by

$$\Theta(e_0, e_1, e_2, e_3) = 1$$

for any oriented Lorentz-orthonormal basis $e_0, \ldots, e_3$ of V. The Hodge star $* : \Lambda^2 V^* \to \Lambda^2 V^*$ and the natural indefinite inner product $\langle,\rangle$ in $\Lambda^2 V^*$ with $\langle u_1 \wedge u_2, u_1 \wedge u_2 \rangle = \langle u_1, u_1 \rangle \langle u_2, u_2 \rangle - \langle u_1, u_2 \rangle^2$ for $u_j \in V^*$, then satisfy

$*^2 = -\mathrm{Id}_{\Lambda^2 V^*}$, $\alpha \wedge *\beta = \langle \alpha, \beta \rangle \Theta$, $\alpha \wedge *\beta = *\alpha \wedge \beta$ and $\langle \alpha, *\beta \rangle = \langle *\alpha, \beta \rangle$ for $\alpha, \beta \in \Lambda^2 V^*$. Regarding $\Lambda^2 V^*$ as a complex 3-space W with $z\alpha = (\mathrm{Re}\, z)\alpha + (\mathrm{Im}\, z) * \alpha$ for $z \in \mathbb{C}$ and $\alpha \in W$, we can define a complex-bilinear (indefinite) inner product $(,)_{\mathbb{C}}$ in W by

$$(\alpha, \beta)_{\mathbb{C}} = \langle \alpha, \beta \rangle - i \langle \alpha, *\beta \rangle.$$

The group $\mathrm{SO}^\uparrow(V)$ acts on W complex-linearly and $(,)_{\mathbb{C}}$-isometrically, so that it also acts on the space $S_0^k W$ of all k-linear symmetric maps $W \times \ldots \times W \to \mathbb{C}$ whose $(,)_{\mathbb{C}}$-contraction is zero. The resulting representation $\mathrm{SO}^\uparrow(V) \to \mathrm{Aut}_{\mathbb{C}}(S_0^k W)$ is irreducible and has the spin invariants $(k, 0)$ (with the actual order depending on the conventions used). Thus, for $k = 1$ we obtain the $(1, 0)$ representation in 2-forms on V and, for $k = 2$, the $(2, 0)$ representation in the space of *algebraic Weyl tensors* on V (see [Besse 1987]).

References: Schweber 1962; Ramond 1981.

3.9 The Choice of Particle Models

The choice of nonrelativistic models of individual particle species in §2.4 was dictated by the fact that to each value of the spin s and parity ϵ there corresponds a unique irreducible representation of $\overline{\mathrm{SU}}(2)$, and hence a unique (first order) natural bundle realizing them. (If $s \notin \mathbb{Z}$, uniqueness is obtained provided a fixed Pauli spinor bundle is used.) This is no more the case in the relativistic category, where a given spin $s = s_1 + s_2$ may be realized by several pairs $\{s_1, s_2\}$ of spin invariants (except for the lowest values $s = 0$ and $s = \frac{1}{2}$ with the unique pairs $\{0, 0\}$, $\{0, \frac{1}{2}\}$) and, moreover, parity is not defined unless $s_1 = s_2$ (§3.7). In contrast with $s = s_1 + s_2$, the individual numbers s_1, s_2 do not seem to represent any invariants measurable by experiment.

Consequently, one obtains the models listed in §1.17 by choosing among the possibilities offered by geometry. More precisely, for bosons (spin $s \in \mathbb{Z}$) one uses the bundle corresponding to the representation with the invariants $s_1 = s_2 = s/2$ and the appropriate parity, while for fermions of spin $s \notin \mathbb{Z}$, the $\{\frac{1}{2}(s + \frac{1}{2}), \frac{1}{2}(s - \frac{1}{2})\}$ representation is selected. (See Examples 3.8.1, 3.8.2.)

The reasons for this choice are quite natural. In both cases, one then achieves a far-reaching formal similarity between the relativistic models and the (unique) nonrelativistic models. Moreover, the parity of bosons remains well-defined, as suggested by experiment.

As for fermion parity, one distinguishes between its two possible (conventional) values by using in §§1.16, 1.17 either a given Dirac spinor bundle, or its complex conjugate, which is based on the fact that the parity of fermions becomes reversed under antiparticle formation, as indicated both by experimental data and the nonrelativistic theory.

We have thus justified the choice of the natural bundle over a given spacetime to represent a particle species with any prescribed spin and parity. As for the corresponding field equations, introduced in §1.16 and §1.17, the reason for selecting them in that particular form can be explained as follows. Suppose that we are working with the Minkowski spacetime M. Any particle of mass m is expected to satisfy the Klein-Gordon equation (1.3), which is the simplest quantized version of the relativistic energy-momentum relation (1.10). However, since quantization of classical formulas (§2.5, §4.8) can lead to loss of information (e.g., quantizing the square of (1.10) instead of (1.10) itself, we obtain a fourth-order equation, much weaker than (1.3)), in some cases the full system of field equations characterizing the particle in question has to be *stronger* than the Klein-Gordon equation alone. For fermions (with spins $s \notin \mathbb{Z}$), the geometry of the bundle provides a natural condition stronger than (1.3) in the form of the Dirac equation (1.7), and so we require that (1.7) be satisfied for all fermions. (Note that (1.7) may fail to imply (1.3) on curved spacetimes, but that does not lead to serious difficulties. See §1.16.)

To see whether and which other conditions have to be imposed, let us fix an "origin" $x_0 \in M$, thus identifying M with its translation space V. The natural bundle η where the given particle lives carries the canonical connection $\overset{\circ}{\nabla}$ (§1.18), which is flat, since so is M. Hence η may be naturally identified with the product bundle $M \times W$, W being the vector space of all $\overset{\circ}{\nabla}$-parallel sections of η. (Thus, $\psi \in \eta_x$ with $x \in M$ corresponds to (x, ϕ), where $\phi \in W$ is the parallel section with $\phi(x) = \psi$.) Sections of η then become just W-valued functions on M. Using the identification $M \cong V$ established by x_0, we can now associate with each section ψ of η, regarded as a function $\psi : V \to W$, its Fourier transform $\hat{\psi} : V^* \to W$, which may also be thought of as a W-valued function on V in view of the isomorphism $V \cong V^*$ due to the Lorentz inner product $\langle , \rangle$. Of course, some assumptions about the growth of ψ have to be made; e.g., ψ may be any W-valued tempered distribution on V. The Klein-Gordon equation (1.3) for any such tempered distribution ψ is equivalent to $(\mathcal{Q} + m^2 c^2 / \hbar^2) \hat{\psi} = 0$ with $\mathcal{Q} : V \to \mathbb{R}$ given by $\mathcal{Q}(v) = \langle v, v \rangle$, which in turn means that $\hat{\psi}$ is the obvious "trivial" extension to V of a distribution on the disconnected 3-manifold $\mathbb{N} = \mathcal{Q}^{-1}(-m^2 c^2 / \hbar^2)$.

Assuming throughout that $m > 0$, let us note that $\mathbb{N}$ then is a two-sheeted hyperboloid in V and either of its components inherits from $\langle , \rangle$ a positive definite Riemannian metric $g_{\mathbb{N}}$ that makes it into a 3-manifold of constant negative sectional curvature, i.e., a hyperbolic space. Let us also assume that the distribution ϕ on $\mathbb{N}$ corresponding to $\hat{\psi}$ is a function $\mathbb{N} \to W$. For instance, any $\phi : \mathbb{N} \to W$ which is measurable and has polynomial growth in terms of V is of this form for a unique W-valued tempered distribution ψ on V. We may then regard ϕ as a section of the vector bundle $\zeta = \mathbb{N} \times W$. Moreover, ζ is natural as a bundle over $(\mathbb{N}, g_{\mathbb{N}})$, since any linear isometry $T_u\mathbb{N} \to T_v\mathbb{N}$ is obtained by restricting a unique linear Lorentz isometry of $V = \mathbb{R}u \oplus T_u\mathbb{N} = \mathbb{R}v \oplus T_v\mathbb{N}$, sending u onto v, and acting on W due to naturality of η. Splitting ζ into irreducible natural subbundles, one obtains summands of spins $s, s-1, \ldots, s-[s]$, each occurring only once, where s is the spin of the particle. (It is so because η has the spin invariants $\{\frac{s}{2}, \frac{s}{2}\}$ or $\{\frac{1}{2}(s + \frac{1}{2}), \frac{1}{2}(s - \frac{1}{2})\}$, cf. §3.4 and the beginning of this section.) It is now quite natural (§2.4) to discard ζ and, instead, use its highest-spin irreducible natural subbundle ζ_s, which has not only the same spin as the particle, but, in the boson case, also the same parity. The condition for $\phi : \mathbb{N} \to W$ to be a section of ζ_s is $\phi(v)(v, \ldots) = 0$ for all $v \in \mathbb{N}$. In fact, $\phi(v) \in W$ is a multilinear, symmetric, zero-contraction map of vectors in V into scalars, pseudoscalars of spinors, and requiring that it be a similar map of vectors in $T_v\mathbb{N}$ amounts to postulating its being "concentrated" in $T_v\mathbb{N} = v^\perp \subset T_v\mathbb{N} \oplus \mathbb{R}v = V$, i.e., that $\phi(v)$ annihilates v. This in turn means that, for the trivial extension $\hat\phi$ of ϕ to V, $\hat\phi(v)(v, \ldots) = 0$ (in the sense of multiplying functions by distributions), so that, for the original section ψ of η, $\operatorname{div}\psi = 0$.

Thus, the additional requirement to be imposed on sections ψ of η turns out to be precisely the divergence condition (1.11). Obviously, that condition is vacuous if $s \leq \frac{1}{2}$, since the number of vector arguments of $\psi(v)$ then is 0 (while $\zeta = \zeta_s$).

References: Ramond 1981; Hayward 1976.

Part II

Unifications

4 The Electromagnetic Interaction

The aim of this chapter is to provide an outline of classical electrodynamics, followed by its reformulation as a gauge theory. The latter approach can in turn be generalized to other interactions (see Chapter 5).

4.1 Nonrelativistic Electrodynamics

Electric and magnetic forces can be regarded as "caused" by two vector fields $\mathbf{E}, \mathbf{B}$ on the Euclidean affine 3-space $\mathbb{M}$, called the *electric* and *magnetic fields*, respectively, and exerting on a test object of electric charge q that moves at a velocity $\dot{\mathbf{x}}$ through a point $\mathbf{x} \in \mathbb{M}$, the force $\mathbf{f}$ given by the *Lorentz force law*

$$\mathbf{f} = q\left[\mathbf{E}(\mathbf{x}) + \dot{\mathbf{x}} \times \mathbf{B}(\mathbf{x})\right]. \tag{4.1}$$

(Since the sign of the vector product $\times$ depends on the choice of an orientation in $\mathbb{M}$, $\mathbf{B}$ is actually a *pseudovector field* on $\mathbb{M}$ and, with the aid of the Hodge star duality, may be identified with an exterior 2-form on $\mathbb{M}$. Cf. Remark 8.7.4.) The fact that the velocity $\dot{\mathbf{x}}$ of the object is affected by the choice of an "inertial observer", i.e., a reference system in $\mathbb{M}$, indicates that $\mathbf{E}$ and $\mathbf{B}$ in general also depend on the observer. This dependence can be clarified with the aid of the relativistic description in §4.2.

The presence of any given electric and magnetic fields $\mathbf{E}, \mathbf{B}$ is usually related to the existence of *sources* that consist of some distribution and motion of electrically charged matter, conveniently described in terms of the electric *charge density* function $\rho : \mathbb{M} \to \mathbb{R}$ and the *velocity* vector field $\mathbf{v}$ of the material continuum. It is also convenient to introduce the *current density* vector field $\mathbf{J} = \rho\mathbf{v}$ on $\mathbb{M}$. Each of $\mathbf{E}, \mathbf{B}, \rho, \mathbf{v}, \mathbf{J}$ may, in general, vary with time. Any prescribed (time-dependent) sources $\rho, \mathbf{J}$ make $\mathbf{E}$ and $\mathbf{B}$ evolve with time t so as to satisfy the *Maxwell equations*

$$
\begin{aligned}
\operatorname{div} \mathbf{E} &= \frac{1}{\epsilon_0}\rho, \\
\operatorname{div} \mathbf{B} &= 0, \\
\operatorname{curl} \mathbf{E} &= -\frac{\partial \mathbf{B}}{\partial t}, \\
\operatorname{curl} \mathbf{B} &= \mu_0\mathbf{J} + \mu_0\epsilon_0\frac{\partial \mathbf{E}}{\partial t},
\end{aligned}
\tag{4.2}
$$

which can be derived using experimental facts, such as the Coulomb and Biot-Savart laws, and heuristic arguments. Maxwell's equations imply the *equation of continuity*

$$\frac{\partial \rho}{\partial t} + \operatorname{div} \mathbf{J} = 0, \tag{4.3}$$

which is interpreted as the conservation law for the electric charge. The universal constants ϵ_0, μ_0 (depending on the units used, and determined by experiment) have the property that, up to measurement errors,

$$\mu_0 \epsilon_0 = \frac{1}{c^2}, \tag{4.4}$$

c being the universal speed of light in the vacuum. Therefore, in regions which are devoid of matter, i.e., such that $\rho = 0$, $\mathbf{J} = 0$ identically, $\mathbf{E}$ and $\mathbf{B}$ satisfy the *wave equation* $\Box \mathbf{E} = \Box \mathbf{B} = 0$ with

$$\Box = -\frac{1}{c^2}\frac{\partial^2}{\partial t^2} + \Delta, \quad \Delta = \sum_j \partial^2/\partial (x^j)^2$$

(cf. §1.16) for Cartesian coordinates x^j in $\mathbb{M}$. Thus, the electric and magnetic fields in the vacuum are vector-valued waves, propagating with the speed c of light. It is now natural to presume that *visible light, and other kinds of similar radiation are electromagnetic waves*, i.e., traveling disturbances of electric and magnetic fields.

Maxwell's equations (4.2) and the Lorentz force law (4.1) together form the foundations of *classical nonrelativistic electrodynamics*, in the sense that they describe how electric and magnetic fields $\mathbf{E}, \mathbf{B}$ have to evolve depending on given sources $\rho, \mathbf{J}$, and, respectively, how they interact with charged matter.

References: Griffiths 1989; Felsager 1983.

4.2 Relativistic Maxwell Equations

By an (*inertial*) *observer* (v_0, y_0) in the Minkowski spacetime M (§1.1) we mean a future-pointing unit timelike vector v_0 in the translation space V of M, regarded as tangent to M at the point $y_0 \in M$. Every fixed observer (v_0, y_0) in M gives rise to the identifications $M \cong \mathbb{R} \times v_0^\perp$, $V \cong \mathbb{R} \times v_0^\perp$, written as $x \sim (x^0, \mathbf{x})$, $w \sim (w^0, \mathbf{w})$ and characterized by $x = y_0 + \mathbf{x} + x^0 v_0$, $w = \mathbf{w} + w^0 v_0$. Here $v_0^\perp$ is the *Euclidean 3-space* of the given observer, i.e., the orthogonal complement of v_0 in V relative to the Lorentz inner product $\langle, \rangle$, corresponding to the metric g of M. Obviously, $\langle w, w \rangle = -(w^0)^2 + |\mathbf{w}|^2$.

Besides the linear function x^0 on M sending each x to $-\langle v_0, x - y_0 \rangle$, one often uses the given observer's *time function* t characterized by $x^0 = ct$, where c is the universal speed of light in the vacuum. Thus, points x of M can be thought of as *events*, identified by each observer with locations $\mathbf{x}$ and times t, so that $x \sim (ct, \mathbf{x})$.

Let us consider an electric field $\mathbf{E}$ and a magnetic field $\mathbf{B}$, with sources $\rho, \mathbf{J}$, all depending on $\mathbf{x} \in \mathbb{M}$ for an affine Euclidean 3-space $\mathbb{M}$, as well as on the time t (§4.1). Choosing an origin in $\mathbb{M}$, we may identify $\mathbb{M}$ and t with the 3-space $v_0^\perp$ of a fixed observer (v_0, y_0) in the Minkowski spacetime M and, respectively, with the observer's time. Then, defining an exterior 2-form F on M and a vector field J on M by

$$F(x)(u, w) = \frac{1}{c} \langle w^0 \mathbf{u} - u^0 \mathbf{w}, \mathbf{E}(\mathbf{x}, t) \rangle + \langle \mathbf{u} \times \mathbf{w}, \mathbf{B}(\mathbf{x}, t) \rangle \qquad (4.5)$$

and

$$J(x) \sim (c\rho(\mathbf{x}, t), \quad \mathbf{J}(\mathbf{x}, t))$$

with $x \in M$, $u, w \in V = T_x M$ and $x \sim (ct, \mathbf{x})$, $u \sim (u^0, \mathbf{u})$, $w \sim (w^0, \mathbf{w})$ for the given observer, we easily see that Maxwell's equations (4.2) can be rewritten as the *relativistic Maxwell equations*

$$\begin{aligned} dF &= 0, \\ \operatorname{div} F &= \mu_0 J, \end{aligned} \qquad (4.6)$$

while the equation of continuity (i.e., conservation of the electric charge) takes the form

$$\operatorname{div} J = 0.$$

Our sign convention is such that, as a vector field on M, the divergence $\operatorname{div} F$ of a 2-form F is characterized by $\langle \operatorname{div} F, u \rangle = \sum_\mu \epsilon_\mu d_{w_\mu}(F(u, w_\mu))$ for constant vector fields u and constant *orthonormal* fields w_μ with $\langle w_\mu, w_\nu \rangle = \epsilon_\mu \delta_{\mu\nu}$, $\epsilon_\mu = \pm 1$, d being the directional differentiation.

The relativisitic Maxwell equations (4.6) are extremely simple and natural in form, and make sense for exterior 2-forms F on any spacetime $(\mathcal{M}, g)$, with the obvious modification of div. It is therefore plausible to postulate that, in the case of the Minkowski spacetime, the *electromagnetic field* F (and hence also the 4-*current* vector field J) be observer-independent. This gives a natural answer to the question how (and "why") $\mathbf{E}$ and $\mathbf{B}$ are affected by a change of the observer (cf. §4.1).

Moreover, one can obviously define an *electromagnetic field on an arbitrary spacetime* $(\mathcal{M}, g)$ to be an exterior 2-form F on $\mathcal{M}$ satisfying (4.6), where J is a given vector field of divergence zero on $\mathcal{M}$, representing the distribution and motion of charged matter.

References: Parrott 1987; Felsager 1983 Dixon 1978; Sachs and Wu 1977.

4.3 Four-momentum and Relativistic Energy

The *world line* of a pointlike object is the timelike C^1 curve in the Minkowski spacetime M, consisting of all points (events) representing the object's presence; see Remark 4.3.1 below. The object's 4-*momentum* at any point x in its world line X then is the unique future-pointing vector p tangent to X at x that satisfies the normalizing condition $\langle p, p \rangle + m^2 c^2 = 0$, m being the mass of the object. For a fixed observer, we have

$$p \sim \left(\frac{E}{c}, \mathbf{p} \right) \tag{4.7}$$

(notation of §4.2), where E and $\mathbf{p}$ are, respectively, called the *total relativistic energy* and the *relativistic 3-momentum* of the object seen by the observer at the time t and location $\mathbf{x}$ with $x \sim (ct, \mathbf{x})$. This terminology is justified as follows. Writing $E = mc^2 + T$, we may think of the observer-independent quantity mc^2 as the *internal* or *rest energy* of the object, and interpret $T \geq 0$ as the *relativistic kinetic energy*, since T and $\mathbf{p}$ satisfy the correct nonrelativistic-limit relations

$$T \to \frac{1}{2} m |\dot{\mathbf{x}}|^2, \ \mathbf{p} \to m\dot{\mathbf{x}} \ \text{ as } \frac{|\dot{\mathbf{x}}|}{c} \to 0.$$

See also Remarks 6.12.5, 6.12.6.

Remark 4.3.1.   Suitable curves in M may be interpreted as trajectories of objects moving in a given observer's 3-space, parametrized by the observer's time (§4.2). Smooth curves that are *timelike* (or, *null*) in the sense that their tangent vectors $v \neq 0$ all satisfy $\langle v, v \rangle < 0$ (or, $\langle v, v \rangle = 0$), clearly correspond to motions at speeds that appear to be less than c (or, equal to c) to *all* observers. The experimental fact that the measured speed c of light in the vacuum is not affected by the motion of the observer, now simply means that the trajectories of light rays are null lines in M. On the other hand, all "normal" objects have to be slower than light (for otherwise they could be slowed down to speed c), and so their world lines are timelike.

Remark 4.3.2.   It is natural to expect that the 4-momentum vector p, with the same properties as above, also exists for objects moving at speed c. However, Remark 4.3.1 then implies that $m^2 c^2 = -\langle p, p \rangle = 0$. Therefore all such objects (including *photons*, the "particles of light"), are *massless*.

References: Dixon 1978; Sachs and Wu 1977; Parrott 1987.

4.4 Relativistic Lorentz Force Law

In contrast with Maxwell's equations (4.6), the Lorentz force law (4.1) is
observer-dependent in the sense that, when interpreted as the equation of
motion of a particle of mass m and charge q (with the aid of *Newton's second
law* $\mathbf{f} = m\ddot{\mathbf{x}}$), it gives rise to different images in the Minkowski spacetime of
the solution trajectories $t \mapsto x(t) \sim (ct, \mathbf{x}(t))$ for different observers (§4.2).
Therefore, the Lorentz force law has to be suitably modified before it can
be regarded as a satisfactory condition in the context of special relativity.
Such a modification is provided by the *relativisitic Lorentz force law*

$$\dot{p} = qF(\cdot, \dot{x}). \tag{4.8}$$

Thus, (4.8) is the equation of motion of a pointlike object of charge q subject
to an electromagnetic field (exterior 2-form) F on M, cf. (4.5). Equation
(4.8) is interpreted as a condition imposed on timelike curves $t \mapsto x(t)$ in M
with $(\ )^{\cdot} = \frac{d}{dt}$, where the parameter t is arbitrary (its choice does not affect
the equation, linear in $(\ )^{\cdot}$), but is usually taken to be a fixed observer's time.
In other words, if t is chosen so that its increasing is compatible with the time
orientation, (4.8) means that $\frac{d}{dt}p(t) = q\,F(x(t))(\cdot, \dot{x}(t))$ with $p(t) = mc\frac{\dot{x}(t)}{|\dot{x}(t)|}$
and $|\dot{x}(t)| = \sqrt{|\langle \dot{x}(t),\ \dot{x}(t)\rangle|}$. (We identify tangent and cotangent vectors
using the Lorentz inner product, so that (4.8) reads $\langle \dot{p}, u \rangle = qF(u, \dot{x})$ for all
vectors u.)

For any fixed observer (§4.2), the 3-space component of (4.8) "becomes"
the ordinary Lorentz force law (4.1) under the limit relation $\mathbf{p} \to m\dot{\mathbf{x}}$ combined with (4.5), while the fourth component of the equality follows from
the other three. The latter fact is no surprise as, due to skew-symmetry of
F, the relativistic Lorentz force law implies that $\frac{d}{dt}\langle p, p \rangle = 0$. This in turn
is compatible with the obvious expectation that the mass m of the object
should remain unchanged, cf. (1.10).

We have thus arrived at the following formulation of *classical relativisitic
electrodynamics*, which makes sense in any spacetime $(\mathcal{M}, g)$. The electromagnetic field is an exterior 2-form F on $\mathcal{M}$, satisfying the Maxwell equations $dF = 0$, $\operatorname{div} F = \mu_0 J$, where J is the 4-current ("source" of F) with
$\operatorname{div} J = 0$. The trajectories (world lines) of pointlike objects of mass m and
charge q are the solutions $t \mapsto x(t) \in \mathcal{M}$, with any parameter t, to the
equation of motion $\dot{p}(t) = F(x(t))(\cdot, \dot{x}(t))$, where $p(t)$ is defined as in spe-
cial relativity, i.e., $p(t) = mc\dot{x}(t)/|\dot{x}(t)|$, as in §4.3, and $(\ )^{\cdot}$ in $\dot{p}$ stands for
the *covariant derivative* $\frac{\nabla}{dt}$.

References: Dixon 1978; Parrott 1987.

4.5 Photons

The 4-current J associated with any electromagnetic field F in a spacetime $(\mathcal{M}, g)$ describes the distribution and motion of charged matter (§4.2, §4.1). Therefore, $J = 0$ identically in any *vacuum region* $U \subset \mathcal{M}$, that is, an open subset consisting of events ("times and locations", for a fixed observer in the case of the Minkowski spacetime) at which no matter is present. Assuming that $U = \mathcal{M}$, i.e., ignoring the remaining part of $\mathcal{M}$, we obtain in $\mathcal{M}$ the electromagnetic field F satisfying the *vacuum Maxwell equations*

$$dF = 0, \quad \operatorname{div} F = 0. \tag{4.9}$$

Numerous experimental facts, such as the photoelectric effect, indicate that under suitable circumstances, electromagnetic radiation behaves like beams of particles, known as *photons*. Photons are not matter fields, but rather carriers of the electromagnetic interaction (§1.5). Therefore, the fact that their classical states (cf. §1.2) are the sections F of the real vector bundle $\eta = \Lambda^2 T^* \mathcal{M}$ is not inconsistent with the usual choice of bundles that represent matter fields (§§1.16, 1.17).

The vacuum Maxwell equations are nothing else than the *field equations for the photon.* If $(\mathcal{M}, g)$ is flat, they imply that

$$\Box F = 0, \tag{4.10}$$

which is the Klein-Gordon equation (1.3) with $m = 0$. Thus, photons are massless, in accord with Remark 4.3.2.

As in the case of Dirac's equation in §1.16, we ignore the difficulties that may arise if g is not flat. In general, (4.9) implies

$$\Box F + \operatorname{Ric} \underset{\widetilde{\ }}{\circ} F + F \circ \operatorname{Ric} + 2R(F) = 0,$$

where F and the Ricci tensor Ric of $(\mathcal{M}, g)$ are regarded as the endomorphisms $v \mapsto \operatorname{Ric}(v, \cdot)$, $v \mapsto F(v, \cdot)$ of $T\mathcal{M}$, while $T\mathcal{M}$ is identified with $T^*\mathcal{M}$ using g, and $R(F)$ stands for the action on F of the curvature operator of $(\mathcal{M}, g)$. For instance, if g has constant curvature K, so that $R(F) = KF$ and $\operatorname{Ric} = 3Kg$, then $\Box F = 4KF$ whenever $dF = 0$ and $\operatorname{div} F = 0$.

Alternative descriptions of the photon can be found in §4.12 and §5.11.

Reference: Parrott 1987.

4.6 Potentials

Let F be an electromagnetic field in a spacetime $(\mathcal{M}, g)$, i.e., a closed exterior 2-form on $\mathcal{M}$ (§4.2). By a *potential* (a *local potential*) for F we mean a 1-form A on $\mathcal{M}$ (or, on an open subset U of $\mathcal{M}$) with $F = dA$. Using g, we often regard A as a vector field on $\mathcal{M}$ (on U).

If $(\mathcal{M}, g)$ is the Minkowski spacetime M, any fixed observer in M allows us to represent a potential A for F by a function $\mathcal{V}$ and a vector field $\mathbf{A}$ on the observer's 3-space, both depending on the observer's time t (§4.2), called a *scalar potential* and a *vector potential* for $\mathbf{E}$ and $\mathbf{B}$ corresponding to F as in (4.5), and characterized by

$$A \sim \left(\frac{\mathcal{V}}{c}, \mathbf{A} \right)$$

A being regarded as a vector field. Relation $F = dA$ now obviously means that $\mathbf{B} = \operatorname{curl} \mathbf{A}$ and $\mathbf{E} = -\nabla \mathcal{V} - \partial \mathbf{A} / \partial t$.

References: Parrott 1987; Felsager 1983; Bogolyubov and Shirkov 1983.

4.7 Lagrangians for a Charged Particle

Let x^{μ}, $\mu = 0, \ldots, 3$, be local coordinates in a spacetime $(\mathcal{M}, g)$. Thinking of $t = x^0/c$ as a time function, cf. §4.2 (which requires assuming that $g(dt, dt) < 0$ and the time orientation of $(\mathcal{M}, g)$ corresponds to increasing t), we may regard x^j, $j = 1, 2, 3$, as "space coordinates". Let us consider a pointlike object of mass m and charge q, which is *free*, i.e., subject to no forces. Its world line then is a timelike geodesic, so that the object is governed by the equation of motion $\dot{p} = 0$, with $(\ )^{\cdot} = \frac{\nabla}{dt}$ (as in §4.4). This is nothing else than the Euler-Lagrange equation for the *free relativistic Lagrangian* $L_{\text{free}} = -mc|\dot{x}|$ for time-parametrized trajectories $t \mapsto x^j(t)$, $j = 1, 2, 3$, which in turn correspond to parametrized world lines $t \mapsto x(t) \in \mathcal{M}$ with $x^0(t) = ct$. The use of L_{free} is very natural for the following reasons. First, the *action functional* it determines associates, with each segment X of a world line, $-mc^2$ times thes *proper time* of X, i.e., the time elapsed along X for the object moving according to X (which is an obvious analogue of the Riemannian curve length). Second, the *canonical momentum components* $p_j = \frac{\partial L}{\partial \dot{x}^j}$ and the *Hamiltonian*, or *total energy* (see [Mackey 1963])

$$H = \dot{x}^j p_j - L \tag{4.11}$$

coincide, for $L = L_{\text{free}}$, with the components of the relativistic 3-momentum, regarded as a 1-form with the aid of g and, respectively, the relativistic total energy, both obtained by decomposing the 4-momentum p as in special relativity, i.e., according to (4.7) with the "infinitesimal observer" $\partial_0/|\partial_0|$ at any fixed point of the coordinate domain (where $\partial_\mu = \partial/\partial x^\mu$).

The Euler-Lagrange equation for any Lagrangian L consists in requiring that the expression

$$\frac{d}{dt}\frac{\partial L}{\partial \dot{x}^j} - \frac{\partial L}{\partial x^j} \tag{4.12}$$

be equal to 0. If an electromagnetic field F is present in the spacetime $(\mathcal{M}, g)$, the free equation of motion $\dot{p} = 0$ (i.e., $\dot{p}_j = 0$) has to be replaced by the Lorentz force law (4.8), i.e., $\dot{p}_j - q\dot{x}^\mu F_{j\mu} = 0$, $j = 1, 2, 3$. (The equality for the 0th components then follows from $\langle p, p \rangle + m^2 c^2 = 0$.) Since (4.12) is linear in L and equals $\dot{p}_j$ for $L = L_{\text{free}}$, we may try to find a Lagrangian for the Lorentz force law in the form

$$L = L_{\text{free}} + L_{\text{int}}$$

with some *interaction term* L_{int} such that (4.12), with L replaced by L_{int}, becomes $-q\dot{x}^\mu F_{j\mu}$. Now, *every* (local) potential A for F leads to $L_{\text{int}} = q\langle A(x), \dot{x} \rangle$ which has the required property since $x^0 = ct$ and $\dot{x}^0 = c$, so

$$L_{\text{int}} = q\dot{x}^\mu A_\mu = cqA_0 + q\dot{x}^k A_k,$$

and hence $\partial L_{\text{int}}/\partial \dot{x}^j = qA_j$, while $F_{\mu\nu} = \partial_\mu A_\nu - \partial_\nu A_\mu$ as $F = dA$ (with $\mu, \nu = 0, \ldots, 3$, $j, k = 1, 2, 3$, and the summing convention).

Combining the canonical momentum $\mathbf{p}_{\text{can}}$ corresponding to $L = L_{\text{free}} + L_{\text{int}}$ with the resulting Hamiltonian $H = E_{\text{can}}$ into the *canonical 4-momentum* vector p_{can} tangent to $\mathcal{M}$ with

$$p_{\text{can}} \sim \left(\frac{H}{c}, \mathbf{p}_{\text{can}} \right) \tag{4.13}$$

(cf. (4.7)), we thus obtain

$$p_{\text{can}} = p_{\text{free}} + qA, \tag{4.14}$$

where p_{free} stands for the ordinary 4-momentum.

The picture thus obtained has an obviously unsatisfactory feature due to the fact that quantities as fundamental as momentum and energy depend on the choice of a potential A for F, which involves enormous arbitrariness. This contradiction can, however, be successfully resolved on the quantum level (see §4.10).

References: Mackey 1963; Emch 1984; Griffiths 1989.

4.8 Relativistic Quantization

Let a classical mechanical system be described in terms of a spacetime $(\mathcal{M}, g)$ and a Lagrangian $L : T\mathcal{M} \to \mathbb{R}$, with x^μ, x^j, t, p_j and H as in §4.7. Its *canonical quantization* then is obtained by selecting a complex vector bundle $\eta \to \mathcal{M}$ with some additional geometric structure that includes a *fixed connection* $\overset{\circ}{\nabla}$ in η, and imposing on sections ψ of η the *field equations*

$$\mathcal{G}(i\hbar \overset{\circ}{\nabla}_{\partial/\partial t}, \ -i\hbar \overset{\circ}{\nabla}_{\partial/\partial x^j}, x^j, t)\psi = 0. \tag{4.15}$$

that arise from the classical energy-momentum relation of the form

$$\mathcal{G}(H, p_j, x^j, t) = 0 \tag{4.16}$$

(given as a part of the system) when the operators $i\hbar \overset{\circ}{\nabla}_{\partial/\partial t}$, $-i\hbar \overset{\circ}{\nabla}_{\partial/\partial x^j}$, $\psi \mapsto x^j \psi$, $\psi \mapsto t\psi$ are substituted for H, p_j, x^j, t in $\mathcal{G}$. This substitution is based on using some (possibly natural, and local) form of functional calculus, and applying the resulting operator to ψ. (See Remark 4.8.1 below.) Of course, only those cases are of interest in which the field equations are independent of the coordinates involved.

Remark 4.8.1. The above quantization procedure is just the canonical quantization of §2.5, rephrased so as to make sense in the relativistic context. Specifically, sections ψ of η may be thought of as evolving versions of L^2 sections of ζ in §2.5 (cf. also §1.2), while the operator substitutions used to obtain (4.15) from (4.16) correspond to the quantization rules for observables, listed in §2.5 and Remark 4.8.4.

Remark 4.8.2. The Klein-Gordon equation (1.3) may be obtained by applying canonical quantization to the classical relation $\langle p, p \rangle + m^2 c^2 = 0$ with $H = E$ (§4.3).

Remark 4.8.3. If $(\mathcal{M}, g)$ is the Riemannian product of $(\mathbb{R}, -c^2 dt^2)$ and $(\mathbb{M}, g_{\mathbb{M}})$, the "nonrelativistic" Schrödinger equation (cf. Remark 4.8.4)

$$i\hbar \frac{\partial \psi}{\partial t} + \frac{\hbar^2}{2m} \Delta \psi = 0$$

for a free particle of mass m can be obtained, through canonical quantization, from the nonrelativistic relation

$$H - \frac{1}{2m}|\mathbf{p}|^2 = 0$$

between the kinetic energy $H = \frac{1}{2}m|\dot{\mathbf{x}}|^2$ and the momentum $\mathbf{p} = m\dot{\mathbf{x}}$. Here ψ is a function, i.e., a section of $\eta = \mathcal{M} \times \mathbb{C}$, and $|\ |$ is the norm corresponding to $g_{\mathbb{M}}$, while Δ is the *Laplacian* of $(\mathbb{M}, g_{\mathbb{M}})$, assigning to a C^2 function ψ on $\mathbb{M}$ its $g_{\mathbb{M}}$-contracted second covariant derivative.

Remark 4.8.4. The evolution of quantum systems with time is governed by the *Schrödinger equation*

$$i\hbar\frac{d\psi}{dt} = H\psi,$$

where H now stands for the total energy observable in the system's Hilbert space $\mathcal{X}$ (§2.5) and $\psi = \psi(t) \in \mathcal{X}$. Relation

$$H = i\hbar\,\partial/\partial t$$

may therefore be regarded as the *quantization rule* for the total energy, analogous to those for position and momentum components, given in §2.5.

Schrödinger's equation can be justified by combining the correspondence principle (§2.5) with Hamilton's evolution equations in classical mechanics.

References: Schweber 1962; Mackey 1963.

4.9 The Minimum-Coupling Substitution

Let a particle of electric charge q interact with a fixed electromagnetic field F on the spacetime $(\mathcal{M}, g)$, represented by a given local potential A (§4.6). Applying canonical quantization as in §4.8, one then expects to end up with a natural vector bundle $\eta \to \mathcal{M}$ (§1.18), the sections of which represent semiclassical states of the particle (§1.2), while the actual choice of η depends on the internal structure of the particle, namely its spin and parity (§§1.16, 1.17). Moreover, sections ψ of η become subject to specific field equations (4.15) obtained by quantizing, as in §4.8, a fundamental classical relation of type (4.16). We expect these field equations to consist of some version of the Klein-Gordon equation (1.3), in most cases with additional conditions such as Dirac's equation (1.7) and/or (1.11), m being the mass of the particle. However, the classical relation $\langle p, p \rangle + m^2 c^2 = 0$, which we obviously must use, is valid for the *ordinary* 4-momentum $p = p_{\text{free}}$, while the quantization procedure leading to the field equations (4.15) consists in using $i\hbar\overset{\circ}{\nabla}_{\partial/\partial t}$ and $-i\hbar\overset{\circ}{\nabla}_{\partial/\partial x^j}$ as quantum versions of H and p_j that, according to (4.13) and (4.14), can be regarded as components of the *canonical* 4-momentum $p_{\text{can}} = p_{\text{free}} + q\,A$. In other words, the classical relation, as well as the corresponding field equation for the particle interacting with the electromagnetic field, are obtained from those valid for a free particle (case $A = 0$) by applying the *minimum-coupling substitution*

$$p_{\text{free}} = p \quad - - - \to \quad p_{\text{free}} = p - qA$$

(with $p = p_{\text{can}}$), or, on the quantum level

$$\overset{\circ}{\nabla}_{\partial/\partial x^\mu} \quad - - - \to \quad \overset{\circ}{\nabla}_{\partial/\partial x^\mu} - \frac{iq}{\hbar} A_\mu \tag{4.17}$$

with $\overset{\circ}{\nabla}$ and the coordinates x^μ in $\mathcal{M}$ as in §4.7. The substitution (4.17), transforming the free field equations into the field equations for a particle subject to the electromagnetic interaction, does not depend on the coordinate system x^μ, as it is equivalent to

$$\overset{\circ}{\nabla}_v \quad - - - \to \quad \overset{\circ}{\nabla}_v - \frac{iq}{\hbar} \langle A, v \rangle \tag{4.18}$$

for all vectors (or vector fields) v tangent to $\mathcal{M}$. In the case where $\eta = \mathcal{M} \times \mathbb{C}$ is the product line bundle with the (flat) product connection $\overset{\circ}{\nabla} = d$, (4.17) consists in replacing $\frac{\partial}{\partial x^\mu}$ by

$$\frac{\partial}{\partial x^\mu} - \frac{iq}{\hbar} A_\mu. \tag{4.19}$$

Here and in (4.17), the functions A_μ are the components of the 1-form $A_\mu dx^\mu$ corresponding to A via g, while the last terms in (4.17) – (4.19) act on sections of η as zero-order operators, i.e., through multiplication.

References: Roman 1961; Felsager 1983; Leite Lopes 1981.

4.10 Potentials as Connection Forms

The problem of having such important quantities as the energy and momentum of a charged particle depend on a highly nonunique (local) potential A, the classical version of which we encountered at the end of §4.7, has obviously reemerged after quantization, as indicated by (4.17) – (4.19). This time, however, we have a very natural way of dealing with it. Namely, an attempt of making the operator (4.19) independent of various possible choices of the A_μ will obviously succeed if one chooses to regard (4.19) as the coordinate expression of a coordinate-independent differential operator, i.e., the *covariant derivative* corresponding to a *connection*. The local components of any connection ∇, here equal to $-\frac{iq}{\hbar} A_\mu$, represent in general a 1-form on the base manifold (which here is $\mathcal{M}$), valued in the Lie algebra $\mathfrak{g}$ of the matrix Lie group G referred to as the *structure group* of the bundle where ∇ lives (cf. §§5.2, 5.3). In our case, $\mathfrak{g} = i\mathbb{R}$ naturally corresponds to the circle $G = \mathrm{U}(1) \cong S^1$, being the tangent line of the unit circle group $S^1 \subset \mathbb{C}$ at 1. Using the standard representation of $\mathrm{U}(1) \cong S^1$ in $\mathbb{C}$, acting by complex multiplication, we can conveniently regard ∇ as a Hermitian (i.e., unitary) connection in a complex line bundle λ over $\mathcal{M}$.

To be specific, we *assume* that there is given an *electromagnetism bundle* λ which represents the electromagnetic interactions involving particles that carry the fixed amount $q \neq 0$ of electric charge. This λ is, by definition, a complex line bundle endowed with a Hermitian fibre metric $\langle , \rangle$ over the spacetime $(\mathcal{M}, g)$ in question. Possible states (configurations) of the electromagnetic field correspond, loosely speaking, to all possible connections ∇ in λ that are *Hermitian*, i.e., compatible with $\langle , \rangle$. With any such connection ∇ and any local unit section ξ of λ we now associate the *local potential A*, which is the 1-form on the domain of ξ, characterized by

$$\nabla \xi = -\frac{iq}{\hbar} A \otimes \xi, \tag{4.20}$$

i.e.,

$$\nabla_v \xi = -\frac{iq}{\hbar} \langle A, v \rangle \xi$$

for all vector fields v. Thus, A is real-valued and

$$\Gamma = -\frac{iq}{\hbar} A \tag{4.21}$$

is the *connection form* of ∇ corresponding to ξ, with

$$\nabla \xi = \Gamma \otimes \xi. \tag{4.22}$$

Moreover, using ξ to identify sections ϕ of λ with functions f in $\mathcal{M}$ (both defined on the domain of ξ), so that $\phi = f\xi$, one easily sees that $\nabla_\mu = \nabla_{\partial/\partial x^\mu}$, acting on sections ϕ, is equivalent to (4.19) acting on the corresponding functions f, for any local coordinate system x^μ in $\mathcal{M}$. The *curvature* R^∇ of ∇, which is a 2-form on $\mathcal{M}$ valued in skew-Hermitian bundle endomorphisms of λ, i.e., in imaginary complex numbers, is given by

$$R^\nabla = -d\Gamma \tag{4.23}$$

for any fixed unit local section ξ, Γ being given by (4.22) (the sign of R^∇ may vary depending on conventions used). Hence, from (4.21) with $dA = F$, we obtain

$$F = -\frac{i\hbar}{q} R^\nabla, \tag{4.24}$$

so that *the electromagnetic field F is proportional to the curvature R^∇ of the connection ∇ we introduced to "replace" F.* Since $dR^\nabla = 0$ (this is the *Bianchi identity*, immediate in our case from (4.23)), Maxwell's equations (4.6) are equivalent to

$$\text{div } R^\nabla = \frac{iq}{\hbar} \mu_0 J \tag{4.25}$$

and, in particular, the vacuum Maxwell equations (4.9) can be rewritten as

$$\text{div } R^\nabla = 0. \tag{4.26}$$

References: Faddeev and Slavnov 1980; Felsager 1983; Leite Lopes 1981.

4.11 Consequences of Introducing the Electromagnetism Bundle

Due to the existence of the electromagnetism bundle λ, postulated in §4.10, the status of the (local) potentials A for a given electromagnetic field F has obviously changed. Instead of being simply all possible (local) "antiderivatives" of F, the potentials A now form a *parametrized family*, the parameters being all pairs (∇, ξ) consisting of a Hermitian connection ∇ in λ with the *prescribed* curvature R^∇ related to F by (4.24), and of any local unit section ξ of λ, with A defined by (4.20) for every such (∇, ξ), so that

$$A(v) = \frac{i\hbar}{q}\langle\nabla_v\xi, \xi\rangle = -\frac{\hbar}{q}\mathrm{Im}\langle\nabla_v\xi, \xi\rangle$$

for each tangent vector v. All local potentials for F are realized in this way, at least locally, even if the connection ∇ with (4.24) is kept fixed. In fact, replacing ξ by $e^{i\varphi}\xi$ with a real-valued function φ, we obtain from (4.20) the analogous formula with $\widetilde\nabla = \nabla$, $\tilde\xi = e^{i\varphi}\xi$ and $\widetilde A = A - \frac{\hbar}{q}d\varphi$. On the other hand, any two potentials for F differ, locally, by the differential of a function. (A undergoes a similar change if, e.g., ξ is fixed and ∇ is transformed using a unitary bundle automorphism of λ. See (5.16).)

The assumption we just made about the existence of the *electromagnetism bundle* λ (for the given charge q), the Hermitian connections ∇ in which are related to possible states F of the electromagnetic field by (4.24), although natural, is in no way an automatic consequence of the previously stated properties of F, namely, the Maxwell equations (4.6). Since the de Rham cohomology class of $\frac{1}{2\pi i}R^\nabla$ in $H^2(\mathcal{M}, \mathbb{R})$ coincides with the image of the first Chern class $c_1(\lambda)$ in $H^2(\mathcal{M}, \mathbb{R})$, for any Hermitian connection ∇ in a complex line bundle λ over any manifold $\mathcal{M}$, the condition that the cohomology class of $\frac{q}{2\pi\hbar}F$ be integral is necessary, by (4.24), for the existence of the required λ and ∇. It is also sufficient, as complex line bundles over $\mathcal{M}$ are classified topologically by $H^2(\mathcal{M}, \mathbb{Z})$ via c_1. Thus, choosing λ such that $c_1(\lambda)$ equals $\frac{q}{2\pi\hbar}F$ in real cohomology and using any fixed Hermitian connection $\widetilde\nabla$ in λ, we see that a given F can be obtained as in (4.24) from the connection $\nabla = \widetilde\nabla - \frac{iq}{\hbar}B$, cf. (4.23), B being a global real 1-form on $\mathcal{M}$ with $F + \frac{i\hbar}{q}R^{\widetilde\nabla} = dB$. On the other hand, if λ exists, it may be nonunique: Topologically, line bundles with a given *real* c_1 are classified by the torsion subgroup of $H^2(\mathcal{M}, \mathbb{Z})$. Moreover, for a fixed F, ∇ is always very far from unique, as $R^{\nabla+idf} = R^\nabla$, by (4.23), for any real-valued function f on $\mathcal{M}$.

Finally, even if λ exists and is topologically unique, it cannot be explicitly constructed using F and the geometry of $(\mathcal{M}, g)$. More precisely, we do not expect λ to be a natural bundle in the sense of §1.18, since neither physics nor geometry gives us any hint as to which natural bundle we should take for λ. Also, if λ were natural, the presence of a preferred connection $\overset{\circ}{\nabla}$ in λ

(§1.18) would restrict the "gauge freedom" (5.16) of choosing Γ (or, A with (4.21)), once $F = dA$ and ξ with (4.22) are fixed.

References: Milnor and Stasheff 1974; Kobayashi and Nomizu 1963.

4.12 Interaction Expressed by Tensor Multiplication

Regarding A in (4.17) , (4.18) as the local potential corresponding to (∇, ξ), where ∇ is a Hermitian connection in the electromagnetism bundle λ and ξ is a local unit section of λ (§4.10), one easily verifies that the minimum-coupling substitution (4.18) can be written as

$$\overset{\circ}{\nabla} \quad ----\to \quad \nabla \otimes \overset{\circ}{\nabla}, \tag{4.27}$$

i.e., amounts to replacing the canonical connection $\overset{\circ}{\nabla}$ in η by the tensor product connection $\nabla \otimes \overset{\circ}{\nabla}$ in the tensor product bundle $\lambda \otimes \eta$. In fact, on the component level, tensor multiplication of connections corresponds to *adding* their components. Specifically, using any fixed unit section ξ of λ over an open set $U \subset \mathcal{M}$ to identify sections ϕ of η with sections ψ of $\lambda \otimes \eta$, over U, by $\psi = \xi \otimes \phi$, we obtain from (4.20)

$$(\nabla \otimes \overset{\circ}{\nabla})_v(\xi \otimes \phi) = (\nabla_v \xi) \otimes \phi + \xi \otimes \overset{\circ}{\nabla}_v \phi = \xi \otimes [\overset{\circ}{\nabla}_v - \frac{iq}{\hbar}\langle A, v\rangle]\phi$$

for any vector field v, so that, under our identification, $(\nabla \otimes \overset{\circ}{\nabla})_v$ becomes the right-hand side of (4.18).

The given particle species of charge q may now be described by two different bundles over the spacetime $(\mathcal{M}, g)$, the *free-particle bundle* η and the *interacting-particle bundle* $\lambda \otimes \eta$, depending on whether we regard it as free (non-interacting, or "ideal"), by ignoring or screening its electric charge, or "switch on" the electromagnetic interaction. According to the minimum-coupling substitution (4.27), the *free field equations* for the particle, imposed on sections ϕ of η and involving the canonical connection $\overset{\circ}{\nabla}$, in order to account for the electromagnetic interaction as well, have to be rewritten in exactly the same form except that ϕ and $\overset{\circ}{\nabla}$ are to be replaced by sections ψ of $\lambda \otimes \eta$ and, respectively, $\nabla \otimes \overset{\circ}{\nabla}$. As a result, one then obtains the corresponding *interacting-particle field equations*. See also §5.4.

On the other hand, Hermitian connections ∇ in λ can be, loosely speaking, identified with semiclassical states of the photon, via (4.24) and §4.5 (cf. §1.2). The field equations for a *free* photon, in the absence of charged matter, are obviously the vacuum Maxwell equations div $R^\nabla = 0$. If charged matter is present, in the form of our particle species of charge q living in the interacting-particle bundle $\lambda \otimes \eta$, we have instead, according to (4.25), the *interacting-photon equations*

$$\text{div } R^\nabla = \frac{iq}{\hbar} \mu_0 J(\psi), \tag{4.28}$$

where the 4-*current* $J = J(\psi)$ is expected to account for the distribution and motion of charged matter and therefore should somehow depend on the state of the charged particle, described by a section ψ of $\lambda \otimes \eta$ (cf. (a) of §1.2). The specific form of this dependence has to be determined, in each case, using additional arguments (see (10.24), (10.25)).

The interacting-particle field equations, obtained via (4.27) from those for the free particle, form along with (4.28) a *coupled* system of equations with unknowns ψ and ∇, describing how states of the given particles and photons evolve with time, while interacting. This system is *nonlinear* (although, as a rule, linear in ψ and ∇ separately), in contrast with the field equations we encountered before, which were all linear (see (1.3), (1.6), (1.7), (1.9), (1.11), (4.6), (4.25)), and often homogeneous. Note that linearity makes sense for connections, since they form an affine space (§5.3).

Consequently, the distinction between linearity and nonlinearity of the field equations can be used as a natural formal criterion of deciding whether the given system of particles is free or interacting.

Reference: Bleecker 1981.

4.13 Quantization of the Electric Charge

One seemingly awkward consequence of having a separate electromagnetism bundle λ (§4.10) *for each value q of the charge* is that one may end up with a family of mutually unrelated line bundles, parametrized by q. In fact, the electromagnetic field F and the potentials A do not depend on the choice of q. Thus, by (4.23) and (4.21), we cannot simply let λ be independent of q, since multiplying the (local) connection form, or the curvature, by a fixed real number, does not correspond to an operation applicable to connections in λ. (The latter operation may also violate the integrality criterion in §4.11 unless λ is topologically trivial.) In other words, once λ is chosen, we are stuck with one particular value of q.

However, a fixed bundle λ corresponding to the charge q naturally gives rise to analogous electromagnetism bundles λ^k representing all integral multiples kq of q, $k \in \mathbb{Z}$. Specifically, we may set

$$\lambda^k = \begin{cases} \bigotimes^k \lambda = \lambda \otimes \ldots \otimes \lambda & \text{if } k \geq 0, \\ \bigotimes^{|k|} \overline{\lambda} & \text{if } k < 0, \end{cases} \tag{4.29}$$

where $\overline{\lambda}$ is the conjugate of λ (§1.6), so that $\overline{\lambda} = \lambda^*$ under the natural isomorphism induced by the fibre metric $\langle,\rangle$. In particular, $\lambda^0 = \mathcal{M} \times \mathbb{C}$, $(\mathcal{M}, g)$ being the spacetime we are working with. For a (local) unit section ξ of λ, we denote by ξ^k the $|k|$-fold tensor power of ξ, regarded as a section of λ^k (note that λ and $\overline{\lambda}$ coincide as real vector bundles). Thus, $\xi^0 = 1$, a constant section of λ^0. Clearly, each λ^k carries a unique Hermitian fibre metric determined by that given in $\lambda = \lambda^1$ in such a way that $|\xi^k| \overset{.}{=} 1$ in λ^k whenever $|\xi| = 1$ in λ. Moreover, every Hermitian connection ∇ in λ gives rise to the Hermitian connection $\nabla^{(k)} = \bigotimes^{|k|} \nabla = \nabla \otimes \ldots \otimes \nabla$, regarded as operating in λ^k. Relations (4.20) – (4.24) then imply

$$\nabla^{(k)} \xi^k = -\frac{ikq}{\hbar} A \otimes \xi^k, \tag{4.30}$$

$$R^{\nabla^{(k)}} = \frac{ikq}{\hbar} F,$$

in view of the Leibniz rule, i.e., the definition of $\bigotimes^{|k|} \nabla$, along with the fact that multiplication by i in λ^k equals multiplication by $-$i in λ^{-k}. Consequently, λ^k plays the same role for kq as λ played for q. Although we have only established (4.30) for connections and unit sections of the form $\nabla^{(k)}, \xi^k$ coming from some ∇, ξ in λ, this is no restriction of generality for connections, as one easily sees using (4.20), while the unit sections in question are local, anyway, and hence may be assumed to have kth roots in λ.

In this way, the electromagnetism bundle λ for the charge q also serves, through its powers λ^k, as an electromagnetism bundle for all multiples kq, $k \in \mathbb{Z}$. As the electromagnetic fields in nature seem to represent a single kind of interactions, it is natural to expect them to have a *unified description*, in terms of just one such bundle λ (and its natural modifications, i.e., powers). Since powers with non-integral exponents cannot be defined in the category of Hermitian complex line bundles, we are thus led to postulate that *there exists an elementary charge q in nature*, such that all possible charge values are its integral multiples. This fact, known as the *quantization of the electric charge*, was first established experimentally by Millikan in 1909, with q that turned out to be the electron charge (or its opposite, the proton charge). There is no contradiction here with the quark charges which are fractional (multiples of $\frac{1}{3}q$, cf. §6.2), since quarks never appear singly, and, moreover, one can always decree the "truly elementary" charge to be $\frac{1}{3}q$.

From now on, unless explicitly stated otherwise, the symbols q and λ will always stand for the electron charge and the corresponding electromagnetism bundle.

Remark 4.13.1. If, in a description that already includes the electromagnetic interaction, a particle lives in a bundle $\lambda^k \otimes \eta$, where $k \in \mathbb{Z}$ and η is natural, or at least does not "involve" λ, we will conclude that the electric charge of the particle equals kq. Consequently, particles represented in this manner by bundles without λ factors ($k = 0$) are electrically neutral.

Remark 4.13.2.   The fact that the electric charge of a composite particle equals the sum of the constituent charges is immediate from Remark 4.13.1 above along with the tensor-product descriptions (1.1), (1.2) of composite objects (cf. (5.3)).

4.14 The Electric Charge and Antiparticles

Let η and $\lambda \otimes \eta$ be the free-particle bundle and the interacting-particle bundle for a particle of charge q, where λ is the electromagnetism bundle for q, as in §4.12. (We do not assume here that q is the elementary charge, and make no use of charge quantization.) Under antiparticle formation (§1.6), these bundles are replaced by $\bar{\eta}$ and

$$\overline{\lambda \otimes \eta} = \overline{\lambda} \otimes \overline{\eta},$$

so that $\overline{\lambda}$ naturally becomes the electromagnetism bundle corresponding to (the charge of) the antiparticle. Since that charge equals $-q$ by (4.29) with $k = -1$, we conclude that *antiparticles have opposite electric charges*.

Reference: Roman 1961.

5 The Yang-Mills Description of Interactions

This chapter contains an exposition of the Yang-Mills approach to interactions, based on representing them with the aid of *gauge fields,* i.e., connections in suitable vector bundles. Such an idea can be justified by the case of electromagnetism, discussed in Chapter 4, where connections appear quite naturally.

5.1 Vector Bundle Operations

To simplify notations, we will often use the ordinary addition and multiplication symbols for the direct sum and tensor product of vector bundles. Thus,

$$\eta_1 + \eta_2 = \eta_1 \oplus \eta_2, \quad \eta_1 \eta_2 = \eta_1 \otimes \eta_2. \tag{5.1}$$

As usual, multiplication will take precedence over addition unless indicated otherwise by parentheses. Equality between bundles will stand for a *natural isomorphism,* which can be formalized as an isomorphism (natural transformation) between functors on an appropriate category. The meaning of naturality, i.e., the choice of the category, may vary with the context and will be specified whenever necessary. For bundles endowed with an additional structure (such as a fibre metric, which may be Riemannian or Hermitian, and possibly indefinite), the sum and product are to be applied to the structures as well, provided that this makes sense in a natural manner. Thus, for instance, all sums of bundles with fibre metrics will be orthogonal.

We will work with bundles over a fixed spacetime $(\mathcal{M}, g)$. With the above operations and the notion of equality just explained, they obviously form a commutative semiring with zero and unity, which are defined by

$$0 = \mathcal{M} \times \{0\}, \quad 1 = \mathcal{M} \times \mathbb{F}, \tag{5.2}$$

$\mathbb{F}$ being the appropriate field ($\mathbb{R}$ or $\mathbb{C}$). We will continue using the notation λ^k for $k \in \mathbb{Z}$ and Hermitian complex line bundles λ, defined by (4.29), since it is consistent with (5.1) for $k \geq 0$. For more general bundles η, $\eta^k = \bigotimes^k \eta$ for $k \geq 0$, according to (5.1), so that $\eta^0 = 1$, while η^{-1} is not defined (as $\bar{\eta}$ would not always be a reasonable substitute). Note that $\eta^{k+\ell} = \eta^k \eta^\ell$ for integers $k, \ell \geq 0$, and, for Hermitian line bundles λ,

$$\lambda^{k+\ell} = \lambda^k \lambda^\ell \tag{5.3}$$

for all $k, \ell \in \mathbb{Z}$. This is immediate since $\lambda^1 \lambda^{-1} = \lambda \overline{\lambda} = 1$ due to the natural isomorphism

$$\lambda_x \overline{\lambda}_x \ni \xi_1 \xi_2 \mapsto \langle \xi_1, \xi_2 \rangle \in \mathbb{C}.$$

Here we have replaced $\otimes$ by the ordinary product symbol not only for bundles, but also for their fibres over $x \in \mathcal{M}$ and individual fibre elements. Thus, we write

$$\psi_1 \dots \psi_k = \psi_1 \otimes \dots \otimes \psi_k \tag{5.4}$$

for sections ψ_j of η_j, $j = 1, \dots, k$.
In particular, we have

$$\Lambda^k(\lambda \eta) = \lambda^k(\Lambda^k \eta) \tag{5.5}$$

for $k \in \mathbb{Z}$ with $k \geq 0$, a Hermitian line bundle λ and any bundle η, due to the natural isomorphism $(\xi_1 \psi_1) \wedge \dots \wedge (\xi_k \psi_k) \mapsto \xi_1 \dots \xi_k (\psi_1 \wedge \dots \wedge \psi_k)$.

5.2 G-structures

Throughout this section, let η be a complex or real vector bundle of fibre dimension N over a manifold $\mathcal{M}$ and let G be a Lie subgroup of the matrix group $GL(N, \mathbb{F})$, where $\mathbb{F}$ is the field of complex or real numbers. Obviously, $GL(N, \mathbb{F})$, and hence also G, acts freely on the set of all bases of any given N-dimensional vector space V over $\mathbb{F}$, including the fibres η_x of η, $x \in \mathcal{M}$. Any orbit of this action of G is called on G-structure in V. By a G-structure in η we mean a selection of one G-structure in each fibre η_x, $x \in \mathcal{M}$, depending smoothly on x in the sense that the union of these G-structures over all $x \in \mathcal{M}$ forms a submanifold of the manifold of all bases of the fibres of η (which in turn is an open subset of the total space of the N-fold direct sum $N\eta = \eta + \dots + \eta$), on which the obvious projection onto $\mathcal{M}$ is a submersion (i.e., a fibre bundle projection). The group G is referred to as the *structure group* of the given G-structure.

For the reader who prefers to use principal bundles, G-structures in η are just principal G-subbundles (i.e., reductions to G) of the principal $GL(N, \mathbb{F})$-bundle of all frames in η, while η is obtained from each of them as the associated bundle for the standard representation of G in $\mathbb{F}^N$.

A G-structure in a vector space V as above gives rise to a group $G(V) \subset \mathrm{Aut}\, V$ of all automorphisms that send some (any) basis in the selected orbit of G onto a basis in the orbit. (Of course, $G(V)$ also depends on the G-structure, i.e., the orbit, but we suppress this from the notation for simplicity.) The actions of $G(V)$ and G on the orbit commute and $G(V)$ is isomorphic to G under an isomorphism $\mathrm{Aut}\, V \cong GL(N, \mathbb{F})$, while no such isomorphism can be canonically distinguished. If a fixed basis in the orbit is used to identify the orbit with G, the actions of $G(V)$ and G on the orbit appear as G acting on itself by the left and, respectively, right translations. Accordingly, the Lie algebra $\mathfrak{g}(V) \subset \operatorname{End} V$ of $G(V)$ is isomorphic to the matrix Lie algebra $\mathfrak{g} \subset \mathfrak{gl}(N, \mathbb{F})$ of G.

Applied to fibres η_x of a bundle η with a fixed G-structure, the above constructions give rise to a bundle $G(\eta)$ of Lie groups and a bundle $\mathfrak{g}(\eta)$ of Lie algebras over $\mathcal{M}$. Sections of $G(\eta)$ of appropriate regularity, depending on the context, are called the *gauge transformations* of the G-structure, and are nothing else than the bundle automorphisms of η leaving the G-structure invariant. The gauge transformations form an infinite dimensional Lie group $\mathcal{G}(\eta)$, to which there naturally corresponds the Lie algebra of *infinitesimal gauge transformations*, consisting of all sufficiently regular sections of $\mathfrak{g}(\eta)$.

G-structures in η are usually equivalent to other types of geometric objects living in the fibres of η, although the specific form of this equivalence is largely a matter of choice and cannot be stated as a general theorem for all matrix groups G. Here are some examples.

(i) For $\mathbb{F} = \mathbb{C}$ and $G = \mathrm{U}(N)$ (or, $\mathbb{F} = \mathbb{R}$ and $G = \mathrm{O}(N)$), a G-structure in η corresponds to a fibre metric $\langle , \rangle$ in η (Hermitian or Riemannian), so that the bases it consists of are precisely the $\langle , \rangle$-orthonormal ones. The fibres of $\mathrm{U}(\eta) = G(\eta)$ and $\mathfrak{u}(\eta) = \mathfrak{g}(\eta)$ (or $\mathrm{O}(\eta) = G(\eta)$ and $\mathfrak{so}(\eta) = \mathfrak{g}(\eta)$) over $x \in \mathcal{M}$ consist of all linear isometries of $(\eta_x, \langle , \rangle_x)$ and of all skew-adjoint endomorphisms of η_x.

(ii) For $G = \mathrm{SL}(N, \mathbb{F})$, a G-structure in η is equivalent to a *volume form* in η, i.e., a nowhere-zero C^∞ section Θ of $\Lambda^N \eta^*$. Bases $\psi_1, \ldots, \psi_N$ forming the G-structure are characterized by

$$\Theta(\psi_1, \ldots, \psi_N) = 1. \qquad (5.6)$$

The fibre of $\mathrm{SL}(\eta) = G(\eta)$ (or $\mathfrak{sl}(\eta) = \mathfrak{g}(\eta)$) over $x \in \mathcal{M}$ consists of all automorphisms of determinant one (or, endomorphisms of trace zero) of η_x.

(iii) For $\mathbb{F} = \mathbb{C}$ and $G = \mathrm{SU}(N)$, a G-structure in η corresponds to a pair $(\langle , \rangle, \Theta)$ of a Hermitian fibre metric $\langle , \rangle$ and a volume form Θ in η (see (ii)) which are *compatible* in the sense that each fibre η_x admits a $\langle , \rangle$-orthonormal basis $\psi_1, \ldots, \psi_N$ with (5.6). It is such bases that then form the G-structure. The fibre of $\mathrm{SU}(\eta) = G(\eta)$ (or, $\mathfrak{su}(\eta) = \mathfrak{g}(\eta)$) over $x \in \mathcal{M}$ consists of all unimodular linear isometries (or, traceless skew-adjoint endomorphisms) of η_x.

(iv) For $G = \{\mathrm{Id}\}$, a G-structure can be identified with a system of trivializing sections $\psi_1, \ldots, \psi_N$ of η.

(v) For $G = \mathrm{GL}(N, \mathbb{F})$, η admits a unique G-structure (which simply corresponds to its vector bundle structure), with $G(\eta) = \mathrm{GL}(\eta) = \operatorname{Aut} \eta$, $\mathfrak{g}(\eta) = \mathfrak{gl}(\eta) = \operatorname{End} \eta$.

Reference: Kobayashi and Nomizu 1963.

5.3 Connections

For $\eta, N, \mathcal{M}$ and G as in §5.2, a (linear) connection ∇ in η is said to be *compatible* with a given G-structure, if ∇-parallel displacements along all curves in $\mathcal{M}$ send bases belonging to the G-structure onto such bases again. ∇ is then called a *G-connection*. In (i) – (iv) of §5.2, compatibility of ∇ is equivalent to the corresponding objects such as $\langle,\rangle$, or Θ, or both, or the ψ_j, being parallel as sections of $\eta^* \otimes \bar{\eta}^*$ (or $\eta^* \otimes \eta^*$), or $\Lambda^N \eta^*$, or η, with respect to the connections ∇ naturally induces there. As for (v) of §5.2, every connection is compatible with the unique $\mathrm{GL}(n, \mathbb{F})$-structure in η, since parallel displacements are linear.

Connections ∇ in η compatible with a given G-structure in η can be identified with sections of a real *affine bundle* (bundle of affine spaces) $\mathcal{C}(\eta)$ whose associated vector bundle, consisting of the corresponding translation spaces, is $\mathfrak{g}(\eta) \otimes T^*\mathcal{M}$. (Again, we omit the dependence on the G-structure in our notation.) In fact, for any two connections ∇, ∇' in η, the difference $\nabla - \nabla'$ is a tensor, namely a section of $(\mathrm{End}\,\eta) \otimes T^*\mathcal{M}$, since ∇ and ∇' are first order differential operators with the same principal symbol. Thus, we may define *connections at a point* $x \in \mathcal{M}$ to be equivalence classes of connections in η modulo the relation $(\nabla - \nabla')(x) = 0$. The connections at x then form an affine space with the translation vector space $(\mathrm{End}\,\eta_x) \otimes T_x^*\mathcal{M}$. Finally, two connections compatible with the same G-structure differ by a section of $\mathfrak{g}(\eta) \otimes T^*\mathcal{M}$, which is immediate from an infinitesimal formulation of compatibility.

References: Milnor and Stasheff 1974; Kobayashi and Nomizu 1963.

5.4 Yang-Mills Fields

In §4.10 and §4.12 we described the electromagnetic interaction, following Hermann Weyl, in terms of compatible connections in a complex line bundle with a $\mathrm{U}(1)$-structure. Since much of that discussion makes sense, at least formally, for vector bundles of any fibre dimension N with a more general G-structure, it is natural to ask if perhaps other types of particle interactions admit similar descriptions, with suitably chosen N and G. This idea of Yang and Mills turned out to be fundamentally correct and became a basic ingredient of currently accepted, and supported by strong experimental evidence, theories of the strong interaction (with $N = 3$, $G = \mathrm{SU}(3)$) and a unified electromagnetic and weak (electroweak) interaction, with $N = 2$ and $G = \mathrm{U}(2)$. See §6.2, §6.5.

Generalizing the discussion of §4.10 and §4.12, let us consider an *interaction bundle* δ, which is a fixed (usually complex) vector bundle of fibre dimension N, endowed with a G-structure for a matrix group $G \subset \mathrm{GL}(N, \mathbb{C})$, over the given spacetime $(\mathcal{M}, g)$. We think of δ as representing some specific interaction, so that, in full analogy with §4.12, we interpret connections ∇ in δ compatible with the G-structure as the semiclassical states of the *interaction carriers* (particles mediating the interaction). These connections, also referred to as *gauge fields* or *Yang-Mills fields*, are sections of the affine bundle $\mathcal{C}(\delta)$ (§5.3), and, if not subject to interactions with other kinds of matter, are expected to satisfy the free field equations

$$\operatorname{div} R^\nabla = 0, \tag{5.7}$$

known as the *Yang-Mills equations*, and obviously generalizing the vacuum Maxwell equations (4.26).

Note that, in general, the curvature R^∇ of ∇ is a 2-form on $\mathcal{M}$ valued in the Lie algebra bundle $\mathfrak{g}(\delta)$, i.e., a section of $\mathfrak{g}(\delta) \otimes \Lambda^2 T^*\mathcal{M}$. To form its divergence, one therefore needs not only the spacetime metric g with its Levi-Civita connection, but also ∇ (i.e., the connection it induces in $\mathfrak{g}(\delta)$). Thus, (5.7) may be *nonlinear* in ∇, since in general R^∇ (or, div R^∇) contains terms quadratic (cubic) in ∇ (cf. (6.39)). In this sense, our interaction carriers may not be "as free" as other free particles (see end of §4.12), which one explains by saying that they display a nontrivial *self-interaction*. Note, however, that R^∇ and (5.7) are linear in ∇ whenever G is Abelian, since the bundle $\mathfrak{g}(\delta)$ then can be naturally trivialized.

On the other hand, let there be given a matter particle capable of taking part in the interaction represented by δ, which we first describe so as to ignore its ability to interact, using the appropriate *free-particle bundle* η over $\mathcal{M}$ (§1.2, §4.12). This natural bundle η carries the canonical connection $\overset{\circ}{\nabla}$ (see §1.18), and is said to host the "free", "ideal", or "generic" version of the given particle species, the choice of terminology depending on circumstances. The particle obeys appropriate free field equations of the form

$$\mathcal{F}(x, \phi(x), (\overset{\circ}{\nabla}\phi)(x), (\overset{\circ}{\nabla}{}^2\phi)(x), \ldots) = 0 \tag{5.8}$$

imposed on its states, i.e., sections ϕ of η, and $x \in \mathcal{M}$.

Let us now "switch on" the interaction described by δ, that is, stop pretending it isn't there. As §4.12 suggests, the no-more free matter particle will now live, instead of η, in the *interacting-particle bundle*, which is *basically* the tensor product $\delta \otimes \eta$ (however, see Remark 5.4.3 below for possible modifications). The way this particle interacts with the interaction carriers is governed, as suggested by §4.12, by the *coupled field equations*

$$\mathcal{F}(x, \psi(x), ((\nabla \otimes \overset{\circ}{\nabla})\psi)(x), ((\nabla \otimes \overset{\circ}{\nabla})^2\psi)(x), \ldots) = 0, \tag{5.9}$$

$$\operatorname{div} R^\nabla = C_0 J(\psi), \tag{5.10}$$

the first of which is obtained from (5.8) via the *minimum-coupling substitution*

$$\overset{\circ}{\nabla} \quad ---\to \quad \nabla \otimes \overset{\circ}{\nabla} \tag{5.11}$$

analogous to (4.28). Equations (5.9), (5.10) are imposed on $x \in \mathcal{M}$ and the states ψ, ∇ of the particles involved, i.e., sections ψ of $\delta \otimes \eta$, and ∇ of $\mathcal{C}(\delta)$. For details, see Remarks 5.4.1, 5.4.2.

Remark 5.4.1. In (5.10), C_0 is a suitable constant (cf. (4.25)), while $J(\psi)$ is a $\mathfrak{g}(\delta)$-valued 1-form on $\mathcal{M}$ (i.e., a section of $T^*\mathcal{M} \otimes \mathfrak{g}(\delta)$) given by an explicit formula involving ψ, whose specific form, for each matter particle and interaction, has to be determined from additional considerations. See Examples 10.3.1 – 10.3.3.

Remark 5.4.2. In all relevant cases, in order to make (5.9) well-defined, $\mathcal{F}$ in (5.8) has to be a function of the variables $x \in \mathcal{M}, \psi \in \eta_x$, $\psi^{(1)} \in T^*_x\mathcal{M} \otimes \eta_x$, $\psi^{(2)} \in T^*_x\mathcal{M} \otimes T^*_x\mathcal{M} \otimes \eta_x, \ldots, \psi^{(k)} \in \bigotimes^k T^*_x\mathcal{M} \otimes \eta_x$, for some k, sending them into the fibre ζ_x of some vector bundle ζ over the given spacetime $(\mathcal{M}, g)$, and making sense for a sufficiently large class of vector bundles η (possibly with some additional structure, and with ζ depending on η). Specifically, the Klein-Gordon equation (1.3), Weyl's equation (1.6), Dirac's equation (1.7), possibly supplemented by the divergence condition (1.11), originally (for the free particle) involving $\overset{\circ}{\nabla}$, can be rewritten using $\nabla \otimes \overset{\circ}{\nabla}$ instead. In fact, this is obvious for (1.3) and (1.11), which make sense for any connection. As for (1.6) and (1.7), the minimum-coupling substitution (5.11) along with the substitution $\mathbf{c} \;- - - \rightarrow\; \mathrm{Id}_\delta \otimes \mathbf{c}$, can be applied to the spinor connection $\overset{\circ}{\nabla}$ in (1.4), thus replacing $\mathcal{D}$ by the *twisted Dirac operator* $\mathcal{D}^\nabla$ in $\delta \otimes \sigma^W$ or $\delta \otimes \sigma$.

Remark 5.4.3. The interaction corresponding to the interaction bundle δ may involve many different species of matter particles, some of them represented by free-particle bundles η' other than η. The interacting-particle bundle then does not have to be $\delta \otimes \eta'$, in contrast with the "most fundamental" case we had for η. For instance, the argument of §4.14 shows that for the antiparticle of the original particle species living in η, with $\eta' = \bar{\eta}$, the interaction bundle has to be $\bar{\delta} \otimes \eta' = \bar{\delta} \otimes \bar{\eta}$. (Thus, as far as antiparticles are concerned, $\bar{\delta}$ is *the* interaction bundle for the given type of interaction.) Furthermore, as illustrated in §4.13 for the electromagnetic interaction, in the case where η' engages in the interaction "less elementarily" than η, the interacting-particle bundle may be of the form $\delta' \otimes \eta'$, where δ' is obtained from δ by a natural (functorial) operation such as $\delta' = \delta^*, \delta' = \bar{\delta}, \delta' = \bigotimes^k \delta, \ \delta' = S^k \delta, \ \delta' = \Lambda^k \delta$, or a combination of these, etc. (which all seem "reasonable" generalizations of the powers λ^k in the case where $\delta = \lambda$ is a Hermitian line bundle). For obvious algebraic reasons, we should also allow interacting-particle bundles of the form

$$\alpha = \delta_1\eta_1 + \ldots + \delta_r\eta_r \tag{5.12}$$

(notation of (5.1)), where $\eta' = \eta_1 + \ldots + \eta_r$ is a natural direct-sum decomposition of η' (we will encounter such a decomposition in (6.4)), and $\delta_1, \ldots, \delta_r$ are obtained from δ by natural operations. The coupled field equations for sections ψ of (5.12) and ∇ of $\mathcal{C}(\delta)$ are again given by (5.9), (5.10), with $\nabla \otimes \overset{\circ}{\nabla}$ interpreted as the connection ∇ and $\overset{\circ}{\nabla}$ naturally induce in the bundle (5.12).

References: Bleecker 1981; Leite Lopes 1981.

5.5 Lack of Naturality

General interaction bundles δ introduced in §5.4 are not natural, for the same reason (or rather lack of it) as at the end of §4.11. Consequently, naturality is usually lost when a free-particle bundle η is replaced by $\delta \otimes \eta$ (or, more generally, by an expression of type (5.12)). This seems to contradict the conclusions of §1.18, and the specific descriptions of bundles that represent matter particles, in §1.16 and §1.17.

The bundle $\mathcal{C}(\delta)$ where interaction carriers live is even worse in this respect, for aside from lack of naturality, it is not even a *vector* bundle. Although we did not insist on using specific models for interaction carriers, the discussion in §1.18 should at least in part apply to them, as long as they are to be treated on a similar footing as other particles.

The following three sections deal with two basic methods of "restoring naturality" in the cases just mentioned.

5.6 Restoring Naturality via Bound States

This method is applicable to interactions whose strength is sufficient to hold several particles together for a detectable length of time, thus forming composite particles known as *bound states* of the original constituents. We are interested in those bound states that are described by natural bundles, so that the particles involved "screen off" or "neutralize" one another's ability to interact with the environment. This screening is never complete in nature, often leaving sizable residual effect (see §1.4), which we are going to disregard. Examples of such bound states living in natural bundles (which can also be referred to as *directly observable, formally stable,* or *formally neutral* bound states) are hadrons, consisting of quarks bound by the strong interaction, and atoms, formed from nuclei and electrons held together by the electromagnetic interaction.

The way formally stable bound states arise is described by (1.1) (or, more generally, (1.2)), where the bundle η representing the composite particle is natural, while $\eta_1, \ldots, \eta_k$ stand for the interacting-particle bundles of the constituents, and the morphism is to be constructed naturally using the available first-order geometry. In other words, the factors in the η_j coming from the interaction bundle δ have to be "killed off", using the geometry of δ in a natural and fibrewise manner. This can only be done with the aid of the objects characterizing the G-structure given in δ, such as $\langle , \rangle$ and/or Θ in (i) – (iii) of §5.2, while the natural (free-particle) factors of the η_j need not be involved, at least at this stage where our only goal is to establish naturality. (This discussion can be made rigorous using simple invariant theory.)

Let $N, G, \langle , \rangle$ and Θ be the fibre dimension of δ (assumed here to be a *complex* vector bundle), its structure group $G \subset \mathrm{GL}(N, \mathbb{C})$, and, respectively, the fibre metric and volume form as in (i) – (iii) of §5.2 (whichever applies for a particular G). Here are a few examples of how (and only how) δ-factors can be "cancelled" in some particular cases:

(i) Let $G = \mathrm{U}(N)$ and let all matter-particle species we are discussing be affected by the interaction "equally strongly" in the sense that each of the interacting-particle bundles is the tensor product of the corresponding free-particle bundle with δ (for species referred to as "particles"), or with $\bar{\delta}$ (for "antiparticles"). The geometry of δ then consists of $\langle , \rangle$ alone, and so the only way of cancelling δ factors is by the use of the natural bundle morphism $\delta\bar{\delta} \to 1$ given by $\xi_1 \xi_2 \mapsto \langle \xi_1, \xi_2 \rangle$ (notation of §5.1). Thus, the only formally stable bound states the interaction δ can form are *particle-antiparticle pairs* (including pairs of completely unrelated species). Note that the analogous $2k$-particle bound states with $k > 1$, given as above by morphisms such as $(\delta\bar{\delta})^k = \delta\bar{\delta} \ldots \delta\bar{\delta} \to 1 \ldots 1 = 1$ do not belong to the category we are dealing with, since such morphisms consist in first sending each of the k factors $\delta\bar{\delta}$ onto 1, *using the interaction* (i.e., $\langle , \rangle$) and then forming a bound state of the resulting k particles living in natural bundles *without involving the interaction* at all. Obviously, we are only interested in the first of these stages.

(ii) Assume that $G = \mathrm{SU}(N)$ and that all matter-particle species involved receive "equal treatment" from the interaction, as in (i). As the geometry of δ now comprises $\langle , \rangle$ and Θ, directly observable bound states formed by the interaction are based on three natural bundle morphisms:

$$\delta\bar{\delta} \to 1, \quad \delta^N \to 1, \quad \bar{\delta}^N \to 1,$$

given by

$$\xi_1 \xi_2 \mapsto \langle \xi_1, \xi_2 \rangle, \; \xi_1 \ldots \xi_N \mapsto \Theta(\xi_1, \ldots, \xi_N), \; \xi_1 \ldots \xi_N \mapsto \overline{\Theta(\xi_1, \ldots, \xi_N)}$$

(notation of (5.1), (5.2), (5.4)). Consequently, the bound states in question consist of *particle-antiparticle pairs, systems of N particles*, and *systems of N antiparticles*. As in (i), in each bound state the constituent (anti)particles may belong to different species. Also, for reasons stated

in (i), we exclude from our consideration further bound states obtained
by iterating the constructions just listed.

(iii) Let $G = \mathrm{U}(N)$ as in (i), but this time allow the strength of the interaction to depend on the matter-particle species involved, in such a way
that each interacting-particle bundle is obtained from the correspond-
ing free-particle bundle by tensoring it with δ^k or $\overline{\delta^k}$ for some integer
$k \geq 0$. The particle then is said to carry the *interaction charge* k or $-k$,
respectively. Setting $\overline{\delta^k} = \delta^{-k}$, we define, for any integers $k_1, \dots, k_n$, a
finite collection of natural bundle morphisms

$$\delta^{k_1} \dots \delta^{k_n} \to \delta^{k_1 + \dots + k_n}, \tag{5.13}$$

obtained by applying the morphism $\delta \overline{\delta} \to 1$ with $\xi_1 \xi_2 \mapsto \langle \xi_1, \xi_2 \rangle$ (notation of §5.1) to selected pairs of factors. (If $N = 1$, this collection
consists of a single isomorphism, cf. (5.3).) As $\delta^0 = 1$ and the geometry of δ consists of $\langle,\rangle$ alone, removing the δ factors from tensor
products of interacting-particle bundles is only possible when the interaction charges $k_1, \dots, k_n$ sum up to zero, and then it has to be done
using the morphisms (5.13). Thus, formally stable bound states can only
be formed from systems of particles with *zero total interaction charge*,
which justifies referring to such states as "neutral".

5.7 Bound States of Interaction Carriers*

Besides systems of matter particles, another possible source of bound states
living in natural bundles are composite particles made of interaction carriers.
However, since interaction carriers are represented by connections ∇ in the
given interaction bundle δ, compatible with its G-structure, i.e., sections
of the *affine* bundle $\mathcal{C}(\delta)$ (§5.4), one cannot even *make sense* of an object
such as $\mathcal{C}(\delta) \otimes \dots \otimes \mathcal{C}(\delta)$, much less map it onto a natural *vector* bundle η
using an explicitly constructed morphism as in (1.1). The best we can do is
try to associate a section u of such an η with any k compatible connections
$\overset{1}{\nabla}, \dots, \overset{k}{\nabla}$, in a manner as close as possible to being k-linear, and algebraic (as
opposed to differential). An obvious idea is to use the curvatures R^j of the $\overset{j}{\nabla}$,
since the dependence of the curvature R^∇ on ∇ is at most quadratic (linear,
if G is Abelian) and always first-order quasi-linear (i.e., R^∇ contains only
first-order derivatives of ∇, and is linear in them), while R^∇ is the simplest
tensorial invariant of ∇. We then naturally obtain the quasi-k-linear map

$$(\overset{1}{\nabla}, \dots, \overset{k}{\nabla}) \mapsto u_k \tag{5.14}$$

assigning to $\overset{1}{\nabla}, \dots, \overset{k}{\nabla}$ the covariant $2k$-tensor u_k on the spacetime $(\mathcal{M}, g)$,
i.e., the section u_k of $\bigotimes^{2k} T^*\mathcal{M}$, with

$$u_k(v_1, \ldots, v_{2k}) = \text{Trace}[R^1_{v_1, v_2} \circ \ldots \circ R^k_{v_{2k-1}, v_{2k}}] \tag{5.15}$$

where $v_j \in T_x\mathcal{M}$ and $R^j_{v,w} = R^j(v, w) \in \mathfrak{g}(\delta_x) \subset \text{End}\,\delta_x$ for $v, w \in T_x\mathcal{M}$, $x \in \mathcal{M}$. Note that when $\overset{1}{\nabla} = \ldots = \overset{k}{\nabla} = \nabla$, the skew-symmetric parts of the u_k contain among them a basis of the characteristic forms of ∇.

One could also form an analogue of (5.15), with the curvatures replaced by *differences* between selected pairs of the connections $\overset{j}{\nabla}$, so that u_k becomes a k-tensor on $\mathcal{M}$, depending on the $\overset{j}{\nabla}$ but not on their derivatives. However, this is not what we need, as the corresponding assignment (5.14) then obviously fails to become k-linear when a fixed connection in δ is used to identify $\mathcal{C}(\delta)$ with the vector bundle $\mathfrak{g}(\delta) \otimes T^*\mathcal{M}$.

It is natural to interpret the mapping (5.14), with (5.15), possibly followed by a projection of u_k onto an irreducible summand of $\bigotimes^{2k} T^*\mathcal{M}$, as a directly observable bound state consisting of k interaction carriers. Since these bound states are *not* matter particles, it is not necessary to restrict our consideration to those irreducible summands of $\bigotimes^{2k} T^*\mathcal{M}$ which are naturally isomorphic to the bundles listed in §1.16 and §1.17. However, as in (1.1), we have to exclude from our discussion those irreducible components of the u_k which vanish identically. (This is the case for u_1 if $G = \text{SU}(N)$ and N is the fibre dimension of δ.)

5.8 Restoring Naturality via Symmetry Breaking

Let δ be an interaction bundle as in §5.4. By *breaking the symmetry* of δ we mean selecting some specific object(s) in each fibre δ_x of δ, which depend smoothly on $x \in \mathcal{M}$ and form at all points x of $\mathcal{M}$ identical configurations with the original geometry (G-structure) of δ_x, up to isomorphisms between fibres. For each particular type of such objects, the corresponding symmetry breaking can be regarded, after appropriate conventions are adopted, as a *reduction* of the original G-structure to a G'-structure contained in it (as a collection of bases in the fibres of η), with some specified subgroup G' of G. Such reductions are not always possible and hence symmetry breaking includes making the corresponding existence assumptions. Letting N be be fibre dimension of δ, one has the following basic examples of symmetry breaking:

(a) A selection of sections $\xi_1, \ldots, \xi_N$ that trivialize δ along with the G-structure, i.e., form at each $x \in \mathcal{M}$ a basis $\xi_1(x), \ldots, \xi_N(x)$ of δ_x belonging to the G-structure. This amounts to a reduction from G to the trivial group $G' = \{\text{Id}\}$;

(b) With $G = \text{U}(N)$, choosing a section ϕ of δ having a nonzero constant length. As a reduction, this corresponds to $G' = \text{U}(N-1)$ embedded in U(N) so as to keep $(1, 0, \ldots, 0) \in \mathbb{C}^N$ fixed, while the G'-structure consists of all orthonormal bases of any fibre δ_x with the fixed first vector $\phi(x)/|\phi(x)|$. (However, fixing the last vector rather than the first, we would get another reduction with a differently embedded U($N - 1$), which illustrates the importance of using specific conventions.)

(c) For $G = $ U(N), selecting a volume form Θ in δ compatible with the Hermitian fibre metric $\langle,\rangle$ corresponding to the U(N)-structure (see (i), (iii) of §5.2). This is equivalent to a reduction from U(N) to SU(N) based on considering, instead of all orthonormal bases $\xi_1, \ldots, \xi_N$ of any δ_x, only those with $\Theta(\xi_1, \ldots, \xi_N) = 1$.

The above definition of symmetry breaking will cover most cases of what physicists refer to as *broken symmetry* in the models we are about to discuss, with the only exception of the SU(5) grand unified theory in §7.1, where the procedure playing the role of symmetry breaking also involves something similar to *choosing a spin structure*. One can easily generalize the definition of symmetry breaking to include such cases by using principal bundles, allowing G' to be just a Lie group with a fixed (not necessarily injective) homomorphism $G' \to G$ and break the symmetry by requiring that there be given a principal G'-bundle over $\mathcal{M}$ along with an equivariant morphism into the principal G-bundle corresponding to the G-structure in δ.

Symmetry breaking can be either *formal* or *physical* (*spontaneous*), depending on whether it reflects any kind of physical reality. Thus, formal symmetry breaking consists in simply *assuming* that the G-structure in δ is reduced (usually to the trivial group), in order to acquire some insight, or interpretation of the structure in question. In nontrivial bundles, this formal game can still be played locally. Since one (local) trivialization of δ along with the G-structure is usually as "good" as any other, we cannot expect such a reduction to have any tangible physical meaning.

On the other hand, spontaneous (physical) symmetry breaking is a geometric description of a class of phenomena that actually take place in nature. For instance, a small sample of water, described nonrelativistically (in an open subset of a Euclidean affine 3-space $\mathbb{M}$) may be thought of as having the structure group O(3), since distances between points in $\mathbb{M}$, and hence lengths of tangent vectors, are well-defined, but no direction (or class of directions, other than the empty or full one) in space is distinguished by the configuration of the sample. However, at low temperatures the sample becomes a small ice crystal, distinguishing several directions of its edges that can be detected using optical effects. Thus, in the process of freezing water, "nature breaks the symmetry" from O(3) to a finite group. A similar effect takes place for some kinds of particle interactions, the role of the low temperature being played there by the relatively low interaction strength (see §6.5.) It should nevertheless be emphasized that, for a geometric description of the ice sample, the use of the resulting finite structure group is not always appropriate. The reason is not only that the sample can melt again but, first of all, the fact that the actual position of the edge directions is completely accidental and the ways it *could have* become still enjoy full $O(3)$ symmetry. Keeping this analogy in mind when discussing the spontaneously broken symmetry for the electroweak interaction in §6.5, we will not simply discard the original structure group $U(2)$, but rather leave it "in the background", aware of the fact that the way nature has "selected" a specific reduction to $U(1)$ (i.e., a nonzero constant-length section of the interaction bundle) from the infinite-dimensional space of a priori possible choices, has been a matter of pure coincidence.

Let us now return to the case of formal symmetry breaking to see how it can help us interpret an interacting-particle bundle $\delta\eta$ or $\overline{\delta}\eta$ (or, more generally, (5.12)) and the affine bundle $\mathcal{C}(\delta)$ for a given interaction bundle δ as certain analogues of models for (usually several) species of matter particles living in natural bundles. To this end we assume we are in the extreme situation where the original structure group G is reduced to the trivial group, i.e., a system of trivializing sections $\xi_1, \ldots, \xi_N$ of δ, compatible with its G-structure, is selected, as in (a) above (we can do it locally, if impossible otherwise). This leads to the natural isomorphism $\delta = \overline{\delta} = 1 + \ldots + 1$ (N copies; notation of §5.1), and hence $\delta\eta = \overline{\delta}\eta = \eta + \ldots + \eta$, while (5.12) becomes a direct sum of the $\eta_1, \ldots, \eta_r$ with possible repetitions. Thus, according to §1.8, the interacting-particle bundle may be regarded as representing several particle species with properties described by the original free-particle bundle (this is why the latter is sometimes said to correspond to a "generic" particle species). On the other hand, $\xi_1, \ldots, \xi_N$ give rise to a unique connection in δ that makes them parallel, which then is automatically compatible with the G-structure. With this selected section, the affine bundle $\mathcal{C}(\delta)$ becomes canonically isomorphic to its translation-space bundle $\mathfrak{g}(\delta)T^*$, where $T^* = T^*\mathcal{M}$. Any fixed basis of the matrix Lie algebra $\mathfrak{g}$ in turn leads to a system of trivializing sections for $\mathfrak{g}(\delta)$, obtained by taking the endomorphisms of the fibres represented in $\xi_1, \ldots, \xi_N$ by matrices of that fixed basis. Thus, naturally, $\mathfrak{g}(\delta) = 1 + \ldots + 1$ (as a real vector bundle, with $1 = \mathcal{M} \times \mathbb{R}$) and $\mathcal{C}(\delta) = \mathfrak{g}(\delta)T^* = T^* + \ldots + T^*$, the number of summands being $\dim G$. Consequently, the interaction carriers may be regarded as consisting of $\dim G$ different species of "real vector bosons", i.e., analogues of matter particles of spin 1 and negative parity which are strictly neutral (see §1.17).

The interpretation just presented is purely formal and should actually be regarded as a figure of speech rather than a description of real particles, due to the fact that the decompositions in question depend heavily on the choice of trivializing sections $\xi_1, \ldots, \xi_N$, which in turn does not reflect anything present in nature, at least in the case of formal symmetry breaking.

References: Bogolyubov and Shirkov 1983; Bleecker 1981; Ryder 1985; Leite Lopes 1981.

5.9 Gauge Symmetry*

Let δ be a vector bundle with a G-structure over the given spacetime $(\mathcal{M}, g)$. The group $\mathcal{G}(\delta)$ of C^∞ gauge transformations of δ (§5.2) then acts, in the obvious manner, not only on sections of δ, but also on G-connections ∇ in δ (§5.3) and their curvatures R^∇. The mapping $\nabla \mapsto R^\nabla$ is $\mathcal{G}(\delta)$-equivariant, which simply expresses naturality of the notion of curvature. (See Remark 5.9.1.)

If G is Abelian, then so is $\mathcal{G}(\delta)$, and hence its action on curvatures is trivial (Remark 5.9.1). This is in particular the case for the electromagnetism bundle λ with $G = U(1)$. However, U(1)-connections ∇ in λ were introduced, for reasons given in §4.10, as a formal replacement of the primary "physical" objects, namely, their curvatures R^∇. In this sense, two U(1)-connections in λ which are *gauge-equivalent,* i.e., congruent under a C^∞ gauge transformation, may be thought of as descriptions of the same physical state, actually represented by their common curvature. Extending this principle to non-Abelian structure groups G, one regards gauge-equivalent connections in a vector bundle δ with a G-structure as *physically equivalent* to an extent that depends on the context. (This is illustrated by Remarks 11.4.1 and 11.5.4. Also, the use of the Lagrangian (10.6) and energy density (10.18) is made more plausible by their gauge invariance.)

However, one cannot simply replace connections ∇ in δ by their gauge-equivalence classes, since the latter would not properly account for interactions with matter, described, e.g., by the substitution (5.11). Instead, it is natural to use gauge-equivalence classes of *pairs* (∇, ϕ) formed by a G-connection ∇ in the interaction bundle δ and a section ϕ of δ. More generally, ϕ may be replaced here by a section ψ of $\delta \otimes \eta$, where η is a given free-particle bundle (§5.4) and each $F \in \mathcal{G}(\delta)$ operates on $\delta \otimes \eta$ as the morphism $F \otimes \mathrm{Id}_\eta$. Since the maps $(\nabla, \phi) \mapsto \nabla\phi$, $(\nabla, \psi) \mapsto (\nabla \otimes \overset{\circ}{\nabla})\psi$ are $\mathcal{G}(\delta)$-equivariant, the interaction in question can still be described in terms of such equivalence classes of pairs, at least under the natural assumption that the field equations (5.9), (5.10) are gauge invariant.

Remark 5.9.1. A gauge transformation F of δ acts on G-connections ∇ in δ according to the formula $(F(\nabla))_v\psi = F(\nabla_v(F^{-1}\psi))$, i.e., $(F(\nabla))_v = \nabla_v - (\nabla_v F)F^{-1}$, or $F(\nabla) = \nabla - (\nabla F)F^{-1}$ for sections ψ of δ and vector fields v on $\mathcal{M}$, where F is regarded as a section of the bundle $\mathrm{End}\,\delta$, which carries a connection naturally induced by ∇. Consequently, $R^{F(\nabla)}(v, w) = F \circ (R^\nabla(v, w)) \circ F^{-1}$ for vector fields v, w on $\mathcal{M}$, and so the tensor fields u_k given by (5.15) with $\overset{1}{\nabla} = \ldots = \overset{k}{\nabla} = \nabla$, as well as the characteristic forms of ∇, are all invariant under gauge transformations acting on ∇.

Remark 5.9.2. In contrast with the curvature invariants mentioned in Remark 5.9.1, the curvature R^∇ itself in general is not left unchanged by gauge transformations of ∇, unless G is Abelian. Since R^∇ is a section of the usually "unnatural" bundle $\mathfrak{g}(\delta) \otimes \Lambda^2 T^*\mathcal{M}$ (§5.4), it is no surprise that one cannot regard it as a directly observable physical quantity.

Remark 5.9.3. A system of sections ξ_α, trivializing δ over an open subset U of $\mathcal{M}$ in a manner compatible with the G-structure, allows us to represent a given gauge transformation $F \in \mathcal{G}(\delta)$ with the aid of the matrix-valued function $\Phi = [\Phi_\alpha^\beta] : U \to G$ such that $F\xi_\alpha = \Phi_\alpha^\beta \xi_\beta$. Furthermore, any G-connection ∇ in δ then can be replaced by the matrix $\Gamma = [\Gamma_\alpha^\beta]$ of connection forms with $\nabla_v \xi_\alpha = \Gamma_\alpha^\beta(v)\xi_\beta$ for vector fields v on $\mathcal{M}$, so that the gauge transformation $\nabla \mapsto F(\nabla)$ corresponds to $\Gamma \mapsto \Phi^{-1}(\Gamma\Phi - d\Phi)$. In particular, if δ is a complex line bundle with $G = U(1)$ (and so $\mathcal{G}(\delta)$ consists of all functions on $\mathcal{M}$ valued in unit complex numbers), we have

$$\tilde{\Gamma} = \Gamma + \mathrm{i}df \tag{5.16}$$

whenever Γ is the connection form of a $U(1)$-connection ∇ in δ relative to a local unit section ξ of δ, in the sense that $\nabla\xi = \Gamma \otimes \xi$ (cf. (4.22)), and $\tilde{\Gamma}$ is given either by $\nabla\tilde{\xi} = \tilde{\Gamma} \otimes \tilde{\xi}$ with $\tilde{\xi} = e^{\mathrm{i}f}\xi$, or by $(F(\nabla))\xi = \tilde{\Gamma} \otimes \xi$, where the gauge transformation F of δ consists in multiplication by $e^{-\mathrm{i}f}$.

Remark 5.9.4. Two G-connections in δ are said to be *locally gauge equivalent* if they become gauge equivalent when restricted to a suitable neighborhood of any point in $\mathcal{M}$. The tensors u_k mentioned in Remark 5.9.1 obviously are invariants of local gauge equivalence. Conversely, in specific situations, some of the u_k may completely characterize local gauge-equivalence classes of G-connections in δ. This is, e.g., the case for u_1 (i.e., the curvature R^∇) if $G = U(1)$ and δ is a complex line bundle. Under those assumptions, R^∇ will also determine ∇ up to a *global* gauge equivalence provided that, in addition, $b_1(\mathcal{M}, \mathbb{Z}) \leq 1$. See §4.10, §4.11.

Remark 5.9.5. Obvious examples of (local) gauge invariants for pairs (∇, ϕ) discussed above are the functions $\langle \phi, \phi \rangle$ and $\langle \phi, \nabla\phi \rangle$, whenever the G-structure in the interaction bundle δ includes a fibre metric $\langle , \rangle$. The same is true for pairs (∇, ψ), where ψ is a section of $\delta \otimes \eta$, provided that the free-particle bundle η carries a fixed fibre metric.

References: Faddeev and Slavnov 1980; Bleecker 1981; Bogolyubov and Shirkov 1983.

5.10 Harmonic Line Bundles*

Let λ be a complex line bundle with a Hermitian fibre metric (i.e., a U(1)-structure, as in (i) of §5.2) over a pseudo-Riemannian manifold $(\mathcal{M}, g)$ of any dimension. A *harmonic structure* in λ is a collection of unit local C^∞ sections of λ whose domains cover $\mathcal{M}$ and whose ratios ("transition functions") are, locally, of the form e^{if} with $\Box f = 0$, where $\Box$ is the Laplacian/d'Alembertian of $(\mathcal{M}, g)$, with $\Box f = \mathrm{div}(df)$ (cf. §1.16) . Any harmonic structure in λ is contained iñ a unique maximal one, and we will say that λ is *harmonic* if it is endowed with a fixed (maximal) harmonic structure. The local unit sections forming this structure then will be referred to as *admissible*. For examples and generalizations, see the remarks below.

For any U(1)-connection ∇ in a harmonic line bundle λ we can define the *divergence* div ∇ of ∇ to be the function div $\Gamma : \mathcal{M} \to i\mathbb{R}$, where Γ is the (imaginary-valued) connection form of ∇ given by (4.22) with any admissible local section ξ. Thus, by (5.16), div ∇ is well-defined (independent of ξ), while (4.23) and the Ricci-Weitzenböck formula for $\psi = \Gamma$ (see, e.g., Remark 10.1.6) imply

$$(\Box - \mathrm{Ric})\nabla = \mathrm{div}\, R^\nabla + d(\mathrm{div}\, \nabla), \tag{5.17}$$

if we define the imaginary-valued 1-form $(\Box - \mathrm{Ric})\nabla$ on $\mathcal{M}$ to be $(\Box - \mathrm{Ric})\Gamma$ with ξ, Γ as above, which again does not depend on ξ. Note that, according to (4.23), R^∇ is imaginary-valued as well.

Remark 5.10.1. One can similarly define harmonic bundles with any Abelian structure group G. However, a straightforward generalization to the non-Abelian case is not possible, since the class of G-valued functions with harmonic logarithms then is not closed under multiplication.

Remark 5.10.2. Given a U(1)-connection ∇ in a harmonic line bundle λ over a pseudo-Riemannian manifold $(\mathcal{M}, g)$ and an imaginary-valued 1-form B on $\mathcal{M}$, one obviously has $\mathrm{div}(\nabla + B) = \mathrm{div}\, \nabla + \mathrm{div}\, B$. Therefore, if $\mathcal{M}$ is compact, the integral

$$\alpha(\lambda) = -i \int_{\mathcal{M}} \mathrm{div}\, \nabla \, dv_g \in \mathbb{R} \tag{5.18}$$

does not depend on the choice of ∇, and so it is an invariant (a "characteristic number") of the harmonic structure of λ.

Remark 5.10.3. In the case where $\lambda = S^1 \times \mathbb{C}$ is the product bundle with the product U(1)-structure over the circle $\mathcal{M} = S^1 = \mathbb{R}/L\mathbb{Z}$ of length $L > 0$, any given C^∞ function $\varphi : S^1 \to \mathbb{R}$ leads to the harmonic structure in λ determined by the collection of admissible sections having the form e^{if} with real-valued functions f defined locally in S^1 and satisfying $f'' = \varphi$. The standard connection $\nabla = d$ in λ then has $\Gamma = i df$ in (4.22) with $\xi = e^{if}$, so that $\operatorname{div} \nabla = i\varphi$, and hence $\alpha(\lambda)$ in (5.18) may assume any prescribed real value, provided that one uses a suitable choice of φ (i.e., of the harmonic structure). The obvious pull-back procedure now also provides examples of harmonic structures with any real value of (5.18) in the product line bundles λ over compact pseudo-Riemannian product manifolds involving S^1 as a factor. In particular, harmonic bundles λ with $\alpha(\lambda) \neq 0$ exist over any flat Riemannian torus T^n.

Remark 5.10.4. Let $(\mathcal{M}, g)$ be an oriented 2-dimensional Riemannian manifold, so that $\mathcal{M}$ can also be regarded as a complex manifold of dimension 1. Every holomorphic line bundle λ over $\mathcal{M}$ with a fixed C^∞ Hermitian fibre metric carries a natural harmonic structure, the admissible sections of which are locally of the form $\psi/|\psi|$, where ψ are local holomorphic sections of λ. The canonical connection ∇ of λ (characterized by $\nabla \psi = 2(\partial \log |\psi|) \otimes \psi$, whenever ψ is holomorphic) then is easily seen to have $\operatorname{div} \nabla = 0$. Thus, by Remark 5.10.3, there exist harmonic line bundles λ over 2-tori that cannot be obtained in this way since, if $\mathcal{M}$ is compact, condition $\alpha(\lambda) = 0$ is obviously *necessary* for λ to be of holomorphic origin. (It is not hard to show that this condition is sufficient as well.)

5.11 The Lorentz Gauge Condition*

An obviously unsatisfactory feature of the description of interactions in terms of connections (§5.4) is that the Yang-Mills equation (5.7) fails to imply anything resembling the Klein-Gordon equation (1.3) or the divergence condition (1.11), despite their otherwise fundamental character. The problem cannot be resolved by simply requiring the field equations to involve additional relations besides (5.7), since (1.3) and (1.11), in general, make no sense when $\psi = \nabla$ is a connection (cf. also §6.9). However, in the case of the electromagnetism bundle λ with the structure group $G = U(1)$, an easy way out of this difficulty can be described as follows. Let us assume that λ is a harmonic bundle (§5.10), so that $\operatorname{div} \nabla$ and $(\square - \operatorname{Ric})\nabla$ are well-defined for U(1)-connections ∇ in λ. We now *declare* the field equations for the free photon to be

$$\operatorname{div} R^\nabla = 0, \quad \operatorname{div} \nabla = 0, \tag{5.19}$$

i.e., to consist of the vacuum Maxwell equation (4.26) and the *Lorentz gauge condition*

$$\operatorname{div} \nabla = 0.$$

By (5.17), these equations imply that $(\Box - \mathrm{Ric})\nabla = 0$, which is a reasonable substitute for the Klein-Gordon equation (1.3), since, for spacetimes $(\mathcal{M}, g)$ with Ric $= 0$, it becomes (1.3) with the correct photon mass $m = 0$ (see §§4.5, 6.9).

The use of the system (5.19) is made plausible by the fact that, physically, it is not more restrictive than Maxwell's equation (4.26) alone. More precisely, in many spacetimes $(\mathcal{M}, g)$, equation $\Box f = \varphi$ is solvable for f with any prescribed (sufficiently regular) function φ and then, by (5.16), every U(1)-connection in λ is gauge equivalent to one with div $\nabla = 0$.

See also Remark 10.1.8.

References: Bogolyubov and Shirkov 1983.

6 The Standard Model

The material covered in this chapter occupies a central position in the classical the theory of particles and interactions. It includes geometric descriptions of the electromagnetic, electroweak and strong (interquark) interactions, based on the unifying Yang-Mills approach outlined in Chapter 5.

6.1 Geometry of Electromagnetism

As already remarked in Chapter 4, electrodynamics may be considered a special case of gauge theory. This approach, proposed by Weyl in 1929, will now be presented using the more general viewpoint described in Chapter 5.

The electromagnetic interaction is a special case of the general interaction scheme in §5.4, corresponding to an interaction bundle λ (the electromagnetism bundle) which is a complex line bundle with a fixed Hermitian fibre metric, i.e., with the structure group $U(1)$, over a given spacetime $(\mathcal{M}, g)$ (see §4.10).

Every particle is endowed with a certain amount of electric charge, which always turns out to be an integral multiple kq, $k \in \mathbb{Z}$, of the charge q carried by the electron, cf. §4.13. (About quarks with their fractional charges, see §4.13 and §6.2.) The interacting-particle bundle for any particle species of charge kq is $\lambda^k \eta$ (notation of (5.1)), where η is the corresponding free-particle bundle. The requirement that a *single* interaction bundle be able to represent, through its powers, i.e., bundles naturally constructed out of it, all electric charges existing in nature, can be regarded as the simplest attempt of a *unification* of particle interactions. Since this amounts to assuming the quantization of the electric charge, the attempt has been successful according to the available experimental evidence.

In view of (iii) of §5.6 with $N = 1$, matter particles with electric charges $k_0 q, \ldots, k_n q$ may form an electromagnetic bound state (composite particle) that lives in a natural bundle, if and only if the system comprising them all is *electrically neutral* in the sense that $k_0 + \ldots + k_n = 0$. Basic examples of such bound states are non-ionized *atoms* of any *atomic number* n, consisting of a *nucleus* with $k_0 = -n$ and of n electrons with $k_1 = \ldots = k_n = 1$. Of course, after the λ factors have been cancelled as in (iii) of §5.6, the remaining tensor product of the free-particle bundles for the nucleus and the n electrons has to be naturally mapped onto an irreducible natural bundle η that represents the atom, thought of as an elementary particle. Some number ℓ of orbital angular momentum units may also be involved, as in (1.2). On the other hand, the required morphism onto η has to be skew-symmetric in the n electron factors, since electrons are fermions (§1.11). This may still leave many possible choices for η and the morphism even if ℓ is fixed, so that the "particles" we obtain represent not just the atom with the given atomic number n, but rather its various possible excitation levels.

Experimental data provide no evidence of any spontaneous symmetry breaking (§5.8) for the electromagnetic interaction that would be valid *generally* in nature, which one expresses by saying that *the* U(1) *symmetry of electromagnetism is exact* (unbroken, perfect). Therefore, we will only apply here the *formal* version of symmetry breaking, with the aim of understanding the particles represented by interacting-particle bundles such as $\lambda^k \eta$ (for the charge kq), and by the affine bundle $\mathcal{C}(\lambda)$. As indicated at the end of §5.8 (with $N = 1$), this leads to identifications $\lambda^k \eta = \eta$, $\mathcal{C}(\lambda) = T^*\mathcal{M}$, i.e., the interacting matter particle appears as the same (single) species as before the interaction was "switched on", while the interaction carrier, living in $\mathcal{C}(\lambda)$, also behaves like a single species, formally similar to a real vector particle of matter, i.e., a strictly neutral particle with spin 1 and negative parity (the *photon*).

More precisely, a fixed, possibly local, unit section ξ of λ identifies compatible connections ∇ in λ with *imaginary-valued* 1-forms Γ on $\mathcal{M}$ by $\nabla \xi = \Gamma \otimes \xi$, so that the photon is a strictly neutral particle with negative C-parity (§1.7).

The only interesting bound state of k photons obtained as in §5.7 from (5.14) and (5.15) corresponds to $k = 1$, since curvatures R^∇ of compatible connections ∇ in λ are 2-forms on $\mathcal{M}$ valued in (imaginary) numbers and so (5.14) with $k > 1$ consists in just putting together sections of natural bundles, without using the electromagnetic interaction at all. However, this is not really a bound state, as only one particle is involved. Instead, one should rather think of $u_1 = R^\nabla$ (with $\nabla = \overset{\downarrow}{\nabla}$, $k = 1$ in (5.14)) as the *directly observable part* ("visible trace") of the photon, which, up to a factor (cf. (4.24)) constituted our previous description of the photon in §4.5.

References: Leite Lopes 1981; Felsager 1983.

6.2 Geometry of the Quark Model

The *quark model*, proposed independently by Gell-Mann and Zweig in 1964 (with subsequent improvements by others), is based on presuming that the strong interaction of hadrons (§1.4), very complicated and defying until then all attempts of a satisfactory description, is a residual effect (see §1.4) of a much stronger but fundamentally simple interaction involving hypothetical constituents of hadrons, known as *quarks*. It is necessary to allow the quarks to come in several "flavors", i.e., species, of which six, denoted u, d, s, c, b, t, are currently believed to exist, along with their antiparticles $\bar{u}, \bar{d}, \bar{s}, \bar{c}, \bar{b}, \bar{t}$, referred to as *antiquarks* (see also Remark 6.2.4). They all have spin $\frac{1}{2}$ and hence are represented by a Dirac spinor bundle σ, assumed to be the same for all quark flavors (§1.16), over the spacetime $(\mathcal{M}, g)$. Thus, the antiquarks all live in the conjugate bundle $\bar{\sigma}$, which happens to be naturally isomorphic to σ (cf. (6.6)). The (conventional) parities are positive for quarks and, consequently, negative for antiquarks (see §1.14).

The most unusual property of quarks to be introduced at this stage is that they have *fractional electric charges*. More precisely, dividing the quarks into three *quark generations*

$$(u, d), \quad (c, s), \quad (t, b), \tag{6.1}$$

one postulates that the electric charges in each generation be $\left(\frac{2}{3}, -\frac{1}{3}\right)$ in units of the proton charge, that is, $\left(-\frac{2}{3}q, \frac{1}{3}q\right)$, where q stands for the charge of the electron. For antiquarks, the charges are opposite (§4.14). This assumption leads to the correct electric-charge predictions for hadrons described as bound states of quarks.

The electromagnetism bundle λ corresponding to the electron charge q obviously cannot be used to describe the electromagnetic interactions involving individual quarks. Instead, one has to assume the existence, and make a choice, of a Hermitian line bundle $\lambda^{1/3}$ along with a fixed identification $(\lambda^{1/3})^3 = \lambda$ (cf. §4.13), as we will do when discussing the SU(5) grand unification in §7.1. On the other hand, since quarks do not seem to appear singly in nature (see Remark 6.2.1 below), for most purposes it will suffice to deal with λ, regarding the electron charge as elementary.

From now on, speaking of the strong interaction, we will always mean the interaction that applies to quarks and *not* its residual form acting upon hadrons (which in our treatment appears as nonexistent). Within the general scheme of §5.4, the strong interaction is described as follows. The *strong-interaction bundle* ρ is a complex vector bundle of fibre dimension 3 over the spacetime $(\mathcal{M}, g)$ and its geometry consists of a Hermitian fibre metric $\langle , \rangle$ and a compatible volume form Θ, i.e., is based on the structure group SU(3) (see (iii) of §5.2). The only matter particles subject to this interaction are quarks and antiquarks, and they enter the interaction in the most basic way, so that the interacting-particle bundle is $\rho\sigma$ for each quark and $\overline{\rho\sigma}$ for each antiquark flavor, σ being the fixed Dirac spinor bundle over $(\mathcal{M}, g)$. Thus, in contrast with electromagnetism, all particles that do interact are subject to the same interaction strength, a "reason" for which is given below in Remark 6.2.2.

The only bound states living in natural bundles that the strong interaction can form out of quarks are thus, by (ii) of §5.6, *quark-antiquark pairs*, *three-quark systems* and *three-antiquark systems*, which one identifies with three disjoint subclasses the set of hadron species falls into, namely, *mesons*, *baryons proper* and *antibaryons*, respectively (§1.12). The quarks contained in a given hadron, along with the way they are put together, which may involve including several units of orbital angular momentum, and then choosing among finitely many possibilities (i.e., irreducible natural summands of the tensor product in (1.2), cf. §8.7), are expected to account somehow for the values of all invariants associated with hadron species.

As far as the *discrete* invariants or quantum numbers (§2.14) are concerned, the agreement is excellent. The lightest hadrons observed in nature coincide exactly with those predicted by the quark model, including their spin, parity, electric charge, isospin, strangeness and other generalized charges (see §8.1 and §8.5 for more details). For heavier hadrons, there are many discrepancies, mainly in the form of hadrons that should be there according to the quark model, but have not yet been observed. It seems reasonable to hope that these difficulties may disappear as further hadrons are discovered, or, perhaps, some simple "exclusion rules" are added to the model.

As an example, here is the quark content of the lightest mesons and baryons listed in the tables of §1.15, ordered by increasing masses, where we omit antiparticles of most particle species that are described, as antiparticle formation applies to each constituent quark. Mesons: $\pi^0(u\bar{u}, d\bar{d})$, $\pi^-(d\bar{u}), \pi^+(u\bar{d}), K^+(u\bar{s}), K^0(d\bar{s}), \eta^0(u\bar{u}, d\bar{d}, s\bar{s}), \rho^-(d\bar{u}), \rho^0(u\bar{u}, d\bar{d}), \rho^+(d\bar{u})$, $\omega^0(u\bar{u}, d\bar{d}), K^{*+}(u\bar{s}), K^{*0}(d\bar{s}), \eta'^0(u\bar{u}, d\bar{d}, s\bar{s}), S^0(u\bar{u}, d\bar{d}, s\bar{s}), \delta^-(d\bar{u}), \delta^+(u\bar{d})$, $\delta^0(u\bar{u},d\bar{d}), \phi^0(s\bar{s}), H^0(u\bar{u},d\bar{d})$. Baryons: $p = N^+(uud), n = N^0(ddu), \Lambda^0(uds)$, $\Sigma^+(uus), \Sigma^0(uds), \Sigma^-(dds), \Delta^-(ddd), \Delta^0(ddu), \Delta^+(uud), \Delta^{++}(uuu)$, $\Xi^-(ssd), \Xi^0(ssu)$. Heavier particles denoted by the same letter as lighter ones are not listed separately, since their quark contents are also the same. Note that most strictly neutral mesons, such as π^0, η^0, ρ^0 etc., have *mixed quark contents* (see (vii) of §8.4).

One easily verifies that the electric charges of the lightest hadrons are properly accounted for by the above quark contents. Note that the charges have to be added when a bound state is formed (Remark 4.13.2).

No spontaneous symmetry breaking (§5.8) for the strong interaction among quarks seems to be taking place in nature, which one expresses by saying that the SU(3) symmetry, or *color symmetry*, of this interaction is *unbroken*. This is not surprising, as it is also the case for the much weaker electromagnetic interaction (§6.1). Therefore, we can only apply here the *formal* version of symmetry breaking, concluding, as at the end of §5.8, that a quark or antiquark subject to the strong interaction, i.e., living in the interacting-particle bundle $\rho\sigma$ or $\overline{\rho\sigma}$, resembles a generalizaton of three particle species, each of which is formally identical to an "ideal" free quark represented by σ (or $\bar{\sigma}$). Similarly, interaction carriers behave to some extent as if they consisted of eight different species of real vector particles, since dim SU(3) = 8. One therefore says that each quark or antiquark flavor comes in three issues, known as *colors*, while the interaction is carried by eight kinds of strictly neutral particles of spin 1 and negative parity, referred to as *gluons*. These notions are mostly formal as any choice of a trivialization of ρ used to arrive at them is as good as any other and hence cannot bear any direct relation to physics. At the same time, the way quarks are divided into colors, and the interaction carriers into gluon species, depends on the trivialization.

Besides the bound states of quarks, corresponding to hadrons, one may also consider bound states of gluons, living in natural bundles, which are given by (5.14), (5.15) (with $k \geq 2$, as $u_1 = 0$ identically whenever the structure group is SU(N)). These bound states are sometimes referred to as *glueballs* and some of them apparently exist in nature, according to indirect experimental evidence.

For more on quarks, see §§7.1, 8.1, 8.3 – 8.7, 9.2 – 9.4.

Remark 6.2.1. Single quarks, with their fractional electric charges, have never been observed, despite frantic searches. One explains that by presuming that the force (interaction) quarks exert on one another is strong enough to prevent them from staying in isolaton for any detectable length of time, before forming bound states (hadrons). For similar reasons, other types of bound states than those named above, such as quark pairs, are impossible. This phenomenon is called *quark confinement* (or *color confinement*) and means, in other words, that the only possible composite states of quarks are the *colorless* ones. By *color* one means here the presence of any non-natural, i.e., ρ-related, factors in the interacting-particle bundle for the composite particle. It is only for the three types of formally stable bound states of quarks, leading to hadrons, that the ρ factors cancel one another, thus rendering the bound state colorless.

Remark 6.2.2. Note that, in contrast with the case of electromagnetism, all particles that do interact strongly, i.e., quarks and antiquarks, are subject to the same interaction strength. The fact that multiple electric charges exist, while "multiple strong changes" do not, can in part be explained as follows. A force *stronger* than electromagnetism, e.g., the residual nuclear interaction, acting on several particles such as protons, along with some electrically neutral neutrons, can hold them together even against their mutual electric repulsion, thus forming nuclei with up to 105 proton charge units. Nothing like this may happen with quarks, leading to "multiply charged" (colored) bound states, simply because there is no force in nature stronger than the strong interaction.

Remark 6.2.3. The mechanism through which quarks determine non-discrete invariant of hadrons, such as the *mass*, seems to be complicated. Nevertheless, there exist good approximate formulae allowing to predict the mass of a hadron based on its quark composition. As for the *quark masses*, the hadron mass spectrum provides no accurate values for them, yet it clearly indicates that they satisfy the inequalities

$$m_u \leq m_d < m_s < m_c < m_b < m_t. \tag{6.2}$$

Remark 6.2.4. All quark flavors other than t (the heaviest) are believed to exist on the basis of indirect experimental evidence, while the existence of the t quark is expected from theoretical considerations. It is plausible that hadrons containing the t quark still escape detection as a consequence of their large masses (see Remark 6.9.5). On the other hand, currently available experimental data make the existence of quarks heavier than t rather unlikely.

References: Ryder 1986, 1985; Leite Lopes 1981; Huang 1982; Sudbery 1986.

6.3 More on Spinors

For our further discussion of the standard model, it is convenient to restate the definition of spinors over a spacetime $(\mathcal{M}, g)$, given in §1.3, as follows (see also Remark 6.3.3 below). Let us assume that $\mathcal{M}$ is orientable, and denote its orientations by L, R (left- or right-handed). By a Weyl spinor bundle over $(\mathcal{M}, g)$ with either fixed orientation L or R, we mean the vector bundle σ_L or σ_R (previously denoted σ^W), associated with a spin structure over $(\mathcal{M}, g)$ via the fundamental representation of Spin $(3, 1)$. Thus, either of σ_L or σ_R is a complex vector bundle of fibre dimension 2 and carries an SL$(2, \mathbb{C})$-structure (cf. §5.2), since Spin $(3, 1)$ = SL$(2, \mathbb{C})$ (see §3.5). Consequently, both σ_L, σ_R are endowed with the corresponding volume forms (see (ii) of §5.2), here denoted ω.

More precisely, SL$(2, \mathbb{C})$ has two "simplest" (lowest-dimensional) non-trivial representations, obtained from each other when the complex structure of $\mathbb{C}^2$ is replaced by its conjugate (§1.6). They are non-equivalent, but become "quasi-equivalent" under an outer automorphism of SL$(2, \mathbb{C})$ that may be chosen as the conjugation by an antiautomorphism of $\mathbb{C}^2$ belonging to a $\mathbb{Z}_2$ extension of SL$(2, \mathbb{C})$ (see (i), (ii) of §3.6). Since such an antiautomorphism projects onto an orientation-reversing element of the Lorentz group $O^\uparrow(3, 1)$, the two different representations correspond to different orientations of $\mathcal{M}$ and hence we have natural, structure-preserving isomorphisms

$$\sigma_R = \overline{\sigma}_L, \quad \sigma_L = \overline{\sigma}_R. \tag{6.3}$$

The corresponding *Dirac spinor bundle* (§1.16) is the direct sum

$$\sigma = \sigma_L + \sigma_R, \tag{6.4}$$

in which the summands are regarded as *unordered*, so that no orientation is distinguished in $\mathcal{M}$. (Obviously, σ also makes sense if $\mathcal{M}$ is nonorientable, the locally defined subbundles σ_L, σ_R then being globally double-valued.)

Moreover, σ carries a naturally defined nowhere-zero section ω of $\Lambda^2\sigma^*$ and a canonical bundle antiautomorphism $(\ )^-$, where ω is obtained by extending the volume forms ω from σ_L, σ_R so that

$$\omega(\sigma_L, \sigma_R) = 0, \tag{6.5}$$

while $(\ )^-$ is the direct sum of the isomorphisms (6.3). Thus, $(\ )^-$ establishes a natural isomorphism

$$\bar{\sigma} = \sigma. \tag{6.6}$$

On each fibre σ_x of σ, the sesquilinear form $\sigma_x \times \sigma_x \ni (\psi, \phi) \mapsto \omega(\psi, \bar{\phi}) \in \mathbb{C}$ is skew-Hermitian and hence the formula

$$\langle \psi, \phi \rangle = i\omega(\psi, \bar{\phi}) \tag{6.7}$$

defines an (indefinite) Hermitian fibre metric in σ.

Our convention about orientations (handedness) is such that the neutrino species ν_e, ν_μ, ν_τ are represented by a fixed left-handed Weyl spinor bundle σ_L, while the antineutrinos live in the corresponding σ_R. All other particles of spin $\frac{1}{2}$ correspond to the Dirac spinor bundle $\sigma = \sigma_L + \sigma_R$ or its conjugate $\bar{\sigma}$, which we keep apart from σ despite the natural isomorphism (6.6). More precisely, σ hosts the leptons proper (e, μ, τ) and all baryons of spin $\frac{1}{2}$ and positive parity (including n, p, Λ), whereas $\bar{\sigma}$ represents the antiparticles of these (the antileptons e^+, μ^+, τ^+, as well as spin $\frac{1}{2}$ baryons of negative parity).

Remark 6.3.1. As we see, most of the natural complex vector bundles η that serve as particle models (§§1.16, 1.17) come equipped with (possibly indefinite) Hermitian fibre metrics $\langle , \rangle$. This includes Dirac spinor bundles σ (cf. (6.7)) and, consequently, all higher-spin fermion models. The only exceptions are the Weyl spinor bundles σ_L, σ_R representing the (anti)neutrinos. (Note that $\langle , \rangle$ in σ, restricted to σ_L or σ_R, is identically zero by (6.3) and (6.5), which in particular implies that the signature of $\langle , \rangle$ is $--++$.)

Remark 6.3.2. It is because of the exceptions mentioned in Remark 6.3.1 that antiparticle formation consists in replacing η by $\bar{\eta}$ rather than the dual bundle η^*, despite the natural isomorphism $\eta^* = \bar{\eta}$ established by $\langle , \rangle$. For Weyl spinor bundles such as σ_L, which represents neutrinos, we have a natural isomorphism $\sigma_L = \sigma_L^*$ established by ω, while σ_L and $\bar{\sigma}_L$ are not naturally isomorphic as they distinguish different orientations of the spacetime. This obviously suggests choosing $\bar{\sigma}_L$ rather than σ_L^* as a model for antineutrinos which, first of all, differ from neutrinos by handedness.

Remark 6.3.3. The above discussion is related to our previous definitions (§1.3 and §1.16) as follows. The underlying Majorana spinor bundle σ^M of the given Dirac spinor bundle σ is the fixed-point set of $(\ )^-$. Then, $(\ )^-$

coincides with the complex conjugation antiautomorphism of $\sigma = \sigma^M + i\sigma^M$ relative to the real form σ^M, i.e., is the unique antilinear extension of Id_{σ^M} to σ. The objects $\pm I$ and ω in σ^M, extended complex (bi)linearly to σ are characterized as $\pm i\cdot\mathrm{Id}$ on σ_L and $\mp i\cdot\mathrm{Id}$ on σ_R (for $\pm I$), and by the definition given earlier in this section (for ω). Conversely, as subbundles of σ, σ_L and σ_R can be described as $\mathrm{Ker}\,(\mathrm{Id} + iI)$ and, respectively, $\mathrm{Ker}\,(\mathrm{Id} - iI)$, while the underlying, mutually conjugate, Weyl spinor bundles for σ^M with the corresponding Ω (or $\overline{\Omega}$) as in §1.3 become naturally isomorphic to σ_L (or σ_R), with ω, under the direct-sum projections $\frac{1}{2}(\mathrm{Id}-iI) : \sigma \to \sigma_L$, $\frac{1}{2}(\mathrm{Id}+iI) : \sigma \to \sigma_R$, restricted to σ^M.

References: Lawson and Michelsohn 1989; Penrose and Rindler 1984–1986; Derdzinski 1991b.

6.4 Hermitian Planes

By a *Hermitian plane* we mean a 2-dimensional complex vector space S with a fixed Hermitian inner product $\langle,\rangle$. We discuss here some properties of Hermitian planes that will be used in a presentation of the electroweak model (§6.5).

Let S be a Hermitian plane. Its *unitary group* $\mathrm{U}(S) \cong \mathrm{U}(2)$ and *unitary Lie algebra* $\mathfrak{u}(S) \cong \mathfrak{u}(2)$ consist of all linear isometries of S and, respectively, all skew-adjoint endomorphisms of S. On $\mathfrak{u}(S)$ one has a 2-dimensional real vector space of (real) bilinear symmetric forms

$$\langle a, b\rangle_{p_0,q_0} = -2p_0\mathrm{Trace}\,(ab) + (p_0 - q_0)\mathrm{Trace}\,a \cdot \mathrm{Trace}\,b \tag{6.8}$$

for $a, b \in \mathfrak{u}(S)$, where p_0, q_0 are real parameters. These forms are naturally distinguished by the Lie algebra structure of $\mathfrak{u}(S)$ (see Remark 6.4.1 below.) We now fix p_0 and q_0, i.e, choose one of the forms (6.8), requiring only that it be positive definite, which is easily seen to be the case if and only if $p_0 > 0$ and $q_0 > 0$. From now on the Euclidean inner product $\langle,\rangle_{p_0,q_0}$ and the resulting norm in $\mathfrak{u}(S)$ will simply be denoted $\langle,\rangle$ and $|\,|$, without any reference to p_0 and q_0, which will lead to no confusion with the inner product and norm in S. Our choice of $\langle,\rangle = \langle,\rangle_{p_0,q_0}$ is, up to a factor, described by the corresponding *Weinberg angle* θ, characterized by

$$\cos^2\theta = \frac{q_0}{p_0 + q_0}, \quad 0 < \theta < \frac{\pi}{2}. \tag{6.9}$$

Thus,

$$\sin^2\theta = \frac{p_0}{p_0 + q_0}, \quad \tan^2\theta = \frac{p_0}{q_0}. \tag{6.10}$$

For an "angular" interpretation of θ, see Remark 6.4.3.

Finally, let us also fix $\phi \in S$ with $\phi \neq 0$. Subspaces of $\mathfrak{u}(S)$ defined in the sequel will, as a rule, *depend on the choice* of p_0, q_0 *and* ϕ, which, for simplicity, will not be reflected in our notation.

As a real Euclidean 4-space, $\mathfrak{u}(S)$ has a direct sum decomposition

$$\mathfrak{u}(S) = \gamma(S) + W(S) + Z(S), \tag{6.11}$$

orthogonal with respect to $\langle,\rangle = \langle,\rangle_{p_0,q_0}$ and determined by the decomposition

$$S = \mathbb{C}\phi + \phi^\perp \tag{6.12}$$

as follows. The real line $\gamma = \gamma(S)$ (or, the plane $W = W(S)$) consists of all $a \in \mathfrak{u}(S)$ with $a\phi = 0$ (or, $a\phi \in \phi^\perp$ and $a(\phi^\perp) \subset \mathbb{C}\phi$), while the line $Z = Z(S)$ is determined by the orthogonality requirement. (About other natural decompositions of $\mathfrak{u}(S)$, see Remark 6.4.2.) The notation we use corresponds to the particle species the summands of (6.11) are related to in the electroweak model (the photon γ, and the W and Z bosons).

Identifying $\mathfrak{u}(S)$ with the matrix Lie algebra $\mathfrak{u}(2)$ with the aid of an orthonormal basis $\phi/|\phi|, \xi$ of S, we see that γ, W and Z correspond, respectively, to the spaces of all matrices

$$\begin{bmatrix} 0 & 0 \\ 0 & r\mathrm{i} \end{bmatrix}, \quad \begin{bmatrix} 0 & -\bar{z} \\ z & 0 \end{bmatrix}, \quad \begin{bmatrix} (p_0+q_0)r\mathrm{i} & 0 \\ 0 & (p_0-q_0)r\mathrm{i} \end{bmatrix}, \tag{6.13}$$

with $r \in \mathbb{R}$, $z \in \mathbb{C}$. Since the squared norms of (6.13) in $\mathfrak{u}(S)$ are, respectively,

$$(p_0+q_0)r^2, \quad 4p_0|z|^2, \quad 4p_0q_0(p_0+q_0)r^2 \tag{6.14}$$

(cf. (6.8)), we obtain the following linear isometries:

$$\gamma(S) \ni a \mapsto (p_0+q_0)^{1/2}\mathrm{Trace}\, a \in \mathrm{i}\mathbb{R}, \tag{6.15}$$

$$W(S) \ni a \mapsto 2p_0^{1/2}|\phi|^{-1}a\phi \in \phi^\perp, \tag{6.16}$$

$$Z(S) \ni a \mapsto p_0^{-1/2}q_0^{1/2}(p_0+q_0)^{1/2}\mathrm{Trace}\, a \in \mathrm{i}\mathbb{R}. \tag{6.17}$$

In particular, using (6.16), we will always regard $W(S)$ as a *complex line*.

For a later application in §11.5, let us also note that

$$a_\gamma \phi = 0, \quad \langle a_w \phi, a_z \phi \rangle = 0,$$
$$2|a_w \phi| = p_0^{-1/2}|\phi|\cdot|a_w|, \quad 2\cos\theta\cdot|a_z\phi| = p_0^{-1/2}|\phi|\cdot|a_z| \tag{6.18}$$

whenever $a_\gamma \in \gamma(S)$, $a_w \in W(S)$, $a_z \in Z(S)$. In fact, if a_γ, a_w or a_z is represented by the corresponding matrix in (6.13), it sends $\phi/|\phi|$ onto the combination of $\phi/|\phi|$, ξ with coefficients given by the first column, so that (6.18) follows from (6.14).

Remark 6.4.1. Formula (6.8) describes all Ad-invariant symmetric bilinear forms on $\mathfrak{u}(S)$, i.e., all biinvariant symmetric 2-tensors on $U(S)$. See also (i) of §6.6.

Remark 6.4.2. The subspace W of $\mathfrak{u}(S)$ and its complement $W^\perp = \gamma + Z$ are naturally determined by ϕ alone (i.e., do not depend on p_0, q_0). On the other hand, even with fixed p_0 and q_0, the plane $W^\perp$ contains other distinguished lines than just γ and Z. There are actually "naturally determined" individual vectors in $W^\perp$, such as i·Id, or the generators of γ and Z obtained from (6.13) with $r = 1$, along with all possible combinations of these. Our choice of γ (and consequently, Z) is motivated by the electroweak model (§6.5).

Remark 6.4.3. Using (6.8) and (6.9) one easily verifies that the Weinberg angle θ equals the angle in $\mathfrak{u}(S)$ (measured using $\langle,\rangle = \langle,\rangle_{p_0,q_0}$) made by the lines γ and $i\mathbb{R}\cdot\mathrm{Id}$.

Remark 6.4.4. Another interpretation of θ comes from the relation

$$\tan\theta = \frac{\operatorname{diam} SU(S)}{\operatorname{diam} S^1} \tag{6.19}$$

(see Remark 6.8.3), where $SU(S)$ and $S^1 = \{z \cdot \mathrm{Id}_S : z \in \mathbb{C}, \ |z| = 1\}$ are the only nontrivial, proper connected normal subgroups of $U(S)$, S^1 being its center, and the diameters are defined using the biinvariant metric on $U(S)$, corresponding to $\langle,\rangle$. See also (i) of §6.6. For either normal subgroup, the diameter equals half the length of any of its one-parameter (circle) subgroups.

6.5 Geometry of the Electroweak Model

After the electromagnetic and strong interactions have been described (§6.1, §6.2) as special cases of the general Yang-Mills interaction model in §5.4, one might expect the same to be possible for the weak interaction. However, this is not the case, for various reasons including the *short-range* character of the weak interaction (see (iii) of §6.6, and Remark 10.3.7). Since carriers of short-range interactions are necessarily massive (Remark 6.9.3), they cannot be directly described in terms of connections (Remark 6.9.4).

The resulting problem can be resolved by unifying the weak and electromagnetic interactions into a single force described by the Yang-Mills formalism and, at the same time, subject to spontaneous symmetry breaking, presumed to actually take place in nature, which reduces its structure group to $U(1)$. As a result, the interaction carriers become split into photons, represented by $U(1)$ connections, and some carriers of the "weak part" of the unified interaction that, for all practical purposes, look like matter particles and so can be (in fact, have to be, cf. §6.9) quite massive. This is the underlying idea of the *electroweak model* that was developed during the 1960s through independent work of Glashow, Salam and Weinberg, reaching its final form in 1967.

All known species of matter particles participate in the weak interaction. However, it is natural to first discuss the electroweak model in its most basic form, where only leptons are affected. An extension of the electroweak theory to other fundamental matter particles, i.e., quarks, leads to additional intricacies, due in part to their fractional electric charges (see §9.4.) The known leptons (§1.14) naturally fall into three categories, known as the *lepton generations*

$$(e, \nu_e), \ (\mu, \nu_\mu), \ (\tau, \nu_\tau), \tag{6.20}$$

so that the electric charges in each generation are $(-1, 0)$ in proton charge units. One does not have to include here the corresponding antiparticles, i.e., antileptons, since each of them is already "accounted for" by the respective lepton generation, cf. (iv) of §6.6.

We will describe the electroweak model for a *single generation* of leptons. A model involving several lepton generations then can be obtained by an obvious direct-sum construction (see (v) of §6.6). For definiteness, we choose the *electron generation* (e, ν_e), containing the electron and the electronic neutrino and, implicitly, also representing their antiparticles e^+ and $\bar{\nu}_e$.

The *electroweak-interaction bundle* ι (cf. §5.4) is a complex vector bundle of fibre dimension 2, endowed with a fixed Hermitian fibre metric $\langle , \rangle$, over the given spacetime $(\mathcal{M}, g)$. Consequently, the geometry of ι is based on the structure group $U(2)$, cf. (i) of §5.2. In physical literature one frequently uses the direct product group $S^1 \times SU(2)$, locally isomorphic to $U(2)$, despite the fact that $U(2)$ is more natural, both geometrically and physically (see (i), (ii) of §6.6).

The free-particle bundle (§5.4) for the electroweak model, more appropriately referred to as the *generic-lepton bundle*, is a fixed Dirac spinor bundle σ over the spacetime $(\mathcal{M}, g)$ (see §6.3). Trying to determine the resulting formally stable bound states would be of little interest (see (vi) of §6.6), so that we will focus on interpreting the corresponding *interacting-lepton bundle* α_{ew} as a generalized-particle model (§1.8) for the lepton species e and ν_e. The actual choice of α_{ew} among the options offered by (5.12) along with the decomposition (6.4), motivated by the knowledge that the subsequent steps of the discussion should lead to e and ν_e described as in (6.26), is

$$\alpha_{ew} = \iota \sigma_L + (\Lambda^2 \iota) \sigma_R \tag{6.21}$$

(notation of (5.1)). Here $\mathcal{M}$ has to be assumed orientable, but not actually oriented, cf. (vii) of §6.6.

The kind of spontaneous symmetry breaking that nature is believed to impose on the original $U(2)$ symmetry should lead to its reduction to $U(1)$, as mentioned before. Therefore, it consists in selecting a C^∞ section ϕ of ι with a *constant* length $|\phi| > 0$ (see (b) of §5.8 and (viii) of §6.6). Such a section always exists provided that $\mathcal{M}$ is noncompact and orientable (by the Euler class argument), and its choice, while breaking the U(2) symmetry, does not obliterate it completely, since nature "could have chosen" any other such section instead (cf. the discussion in §5.8). The reason why ϕ and not a different structure is selected in ι can also be justified from dynamical considerations without invoking the goal of a reduction to U(1) (Remark 11.5.2).

With ϕ chosen as above, the complex line subbundle

$$\lambda = \phi^\perp \tag{6.22}$$

of ι obviously inherits a Hermitian fibre metric, i.e., a U(1)-structure, from ι. We call λ *the electromagnetism bundle* and regard it as corresponding to the electric charge q carried by the electron (§4.10, §6.1). The natural, isometric bundle isomorphisms $\iota = \mathbb{C}\phi + \phi^\perp$, i.e.,

$$\iota = 1 + \lambda \tag{6.23}$$

(notation of (5.1), (5.2)) and

$$\lambda = \Lambda^2 \iota, \tag{6.24}$$

the latter defined by

$$\lambda_x = \phi(x)^\perp \ni \xi \mapsto \xi \wedge \frac{\phi(x)}{|\phi(x)|} \in \Lambda^2 \iota_x, \ x \in \mathcal{M}, \tag{6.25}$$

along with (6.4) and (6.21), yield

$$\alpha_{\mathrm{ew}} = \sigma_{\mathrm{L}} + \lambda\sigma. \tag{6.26}$$

This is exactly the desired picture of the interacting-lepton bundle α_{ew}, now generalizing the neutrino ν_{e}, living in σ_{L}, and the electron e, represented by $\lambda\sigma$. In fact, the presence or absence of the factor λ indicate the correct amounts of the electric charge for e and ν_{e} (see Remark 4.13.1), while the free-particle bundles, σ_{L} for ν_{e} and σ for e, coincide with those dictated by the handedness and spin of either particle (§1.16, §6.3).

Before we proceed to discuss the interaction carriers, the following remark is in order. Vector bundles representing elementary particles are expected to carry, whenever possible, Hermitian or Riemannian fibre metrics that are natural, but not necessarily definite, while all relevant direct-sum decompositions of such bundles should be orthogonal (Remark 6.3.1). Exceptions to this rule, forced upon us due to their being "very natural" in other ways, are the Weyl spinor bundles $\sigma_{\mathrm{L}}, \sigma_{\mathrm{R}}$ (and hence also α_{ew}), as well as the decompositions (6.4), (6.26). The above general requirement obviously makes sense for affine bundles as well, with a translation-invariant metric in each fibre, so that a fibre metric in an affine bundle is just an ordinary fibre metric in the associated vector bundle of translation spaces.

This can be applied, in particular, to the real affine bundle $\mathcal{C}(\iota)$ representing the electroweak interaction carriers, the sections of which are just connections in ι compatible with $\langle , \rangle$ (i.e., with the U(2)-structure, §5.3). The translation-space bundle for $\mathcal{C}(\iota)$ is the tensor product $\mathfrak{u}(\iota)T^*$ with $T^* = T^*\mathcal{M}$, and natural Riemannian fibre metrics in it can be obtained by tensoring the metric in T^* induced by g with any of the positive definite fibre metrics in $\mathfrak{u}(\iota)$ given, for fixed positive numbers p_0, q_0, by (6.8) in each fibre $S = \iota_x$ of ι. Let us choose such p_0 and q_0 and from now on always use the corresponding fibre metrics in $\mathfrak{u}(\iota)$, $\mathfrak{u}(\iota)T^*$ and $\mathcal{C}(\iota)$, along with the angle θ defined by (6.9), known as the *Weinberg angle* (or *Glashow–Weinberg angle*) of the electroweak model.

To determine the electroweak-interaction carriers, let us note that compatible connections in ι making ϕ parallel coincide with the sections of a specific affine subbundle $\mathcal{C}_\phi(\iota)$ of $\mathcal{C}(\iota)$. Moreover, applying (6.11) to each fibre $S = \iota_x$ of ι with $\phi = \phi(x)$ and the fibre metric in $\mathfrak{u}(\iota)$ corresponding to our fixed p_0, q_0, we obtain the orthogonal direct-sum decomposition of vector bundles $\mathfrak{u}(\iota) = \gamma(\iota) + W(\iota) + Z(\iota)$. Since the translation-space bundle of $\mathcal{C}_\phi(\iota)$ is the subbundle $\gamma(\iota)T^*$ of $\mathfrak{u}(\iota)T^*$, due to the definition of $\gamma(S)$ in §6.4, this implies the orthogonal affine-bundle decomposition

$$\mathcal{C}(\iota) = \mathcal{C}_\phi(\iota) + W(\iota)T^* + Z(\iota)T^*. \tag{6.27}$$

(cf. (x) of §6.6). We also have the natural isomorphism

$$\mathcal{C}_\phi(\iota) = \mathcal{C}(\lambda), \tag{6.28}$$

obtained by restricting to λ the connections in ι that make ϕ parallel, and hence leave $\lambda = \phi^\perp$ invariant. The associated isomorphism $\gamma(\iota)T^* \to T^*$ of vector bundles equals $-i\,\mathrm{Trace} \otimes \mathrm{Id}_{T^*}$ and, as (6.15) is isometric, (6.28) will become isometric provided that the fibre metric $\langle , \rangle_g$ in $\mathcal{C}(\lambda)$, induced by g in T^*, is replaced by

$$\langle , \rangle = (p_0 + q_0)\langle , \rangle_g. \tag{6.29}$$

On the other hand, applying (6.16), (6.17) to each fibre $S = \iota_x$ of ι we obtain isometric bundle isomorphisms $W(\iota) = \phi^\perp = \lambda$, $Z(\iota) = 1 = \mathcal{M} \times \mathbb{R}$, so that (6.27) and (6.28) give rise to the natural bundle isomorphism

$$\mathcal{C}(\iota) = \mathcal{C}(\lambda) + \lambda T^* + T^*, \tag{6.30}$$

preserving the fibre metrics (except for a factor of $p_0 + q_0$ in the case of $\mathcal{C}(\lambda)$; see also Remark 6.8.1).

Consequently, the electroweak interaction carriers, living in $\mathcal{C}(\iota)$, form a generalization (§1.8) of the following particle species, corresponding to the summands of (6.30):

(i) The photon γ, carrier of the electromagnetic interaction, represented by sections of $\mathcal{C}(\lambda)$, i.e., U(1)-connections in λ, exactly as in §6.1.

(ii) Two species of weak-interaction carriers, known as the W boson and the Z boson, living in the summands λT^* and T^*, respectively, and hence appearing, formally, as if they were matter particles, both of spin 1 and negative parity (§1.17). More precisely,

a) As the factor λ indicates, W carries an electric charge, equal to that of the electron (Remark 4.13.1), and so we write $W = W^-$. Consequently, W^- is not strictly neutral (§1.6). In fact, the real tensor product λT^* is a *complex* bundle, since so is λ, and we have the natural complex vector-bundle isomorphism

$$\lambda T^* = \lambda \underset{\mathbb{R}}{\otimes} T^* = \lambda \underset{\mathbb{C}}{\otimes} (T^* \otimes \mathbb{C}).$$

As an "ideal" free particle, W^- thus lives in the complexification $T^* \otimes \mathbb{C}$ of $T^* = T^*\mathcal{M}$, in agreement with §1.17.

b) The antiparticle W^+ of W^-, with the opposite electric charge, is another carrier of the weak interaction, which we regard as already accounted for, in the above discussion, by λT^* (when charged) or $T^* \otimes \mathbb{C}$ (if its charge is ignored). Cf. (iv) of §6.6.

c) The Z boson is represented by the *real* bundle $T^* = T^*\mathcal{M}$, and hence it is strictly neutral (§1.6), in particular electrically neutral: $Z = Z^0$. (Due to the factor of i in (6.17), Z^0 has negative C-parity, cf. §1.7.)

The W and Z bosons were discovered experimentally in 1983, almost a decade after the electroweak model, confirmed by plentiful experimental evidence, as well as by its compatibility with the field quantization procedure, became an established part of physical theory. The large masses of W and Z made their detection impossible before the advent of sufficiently high-energy accelerators (see Remarks 6.9.1, 6.9.5).

It is worth noting that the prediction of the Z boson was one of the reasons why the electroweak model originally met with skepticism. In contrast with the charged weak-interaction carriers (the $W^\pm$ bosons), the existence of Z^0 was not supported at that time even by indirect experimental evidence in the form of reactions that a neutral weak-interaction carrier could mediate. The discovery of such reactions, known as *weak neutral currents*, in 1973 constituted a major step towards general acceptance of the electroweak theory.

The way neutrinos appear in the electroweak model is consistent with the currently prevailing opinion that they are massless. However, should experimental evidence indicate that this is not the case (which is also suggested by some concepts in cosmology), the electroweak model can easily be modified, actually becoming even simpler, so as to accomodate massive neutrinos. See Remark 6.9.2.

Although the Weinberg angle θ is a free parameter of the theory, its choice is not arbitrary. Many predictions of the electroweak model depend on θ and, at the same time, can be verified by experiment. The resulting data consistently lead to the formula

$$\sin^2\theta = 0.234 \pm 0.013 \tag{6.31}$$

characterizing an approximate value of the Weinberg angle in nature. This θ along with the electron charge q, also determined experimentally, lead via (6.10) and (6.37) to unique "experimental" values of p_0 and q_0.

For more on the electroweak model, see §§6.6, 6.8, 6.9, 9.4 and 11.5.

References: Ryder 1985, 1986; Leite Lopes 1981; Huang 1982; Sudbery 1986.

6.6 Remarks

(i) The N-fold covering homomorphism $S^1 \times SU(N) \to U(N)$ sending (z, F) onto zF corresponds to a Lie algebra isomorphism $\mathbb{R} \oplus \mathfrak{su}(N) \cong \mathfrak{u}(N)$ and, for $N \geq 2$, gives rise to the only nontrivial, proper, connected normal subgroups $S^1 = \{z \cdot \mathrm{Id} : z \in \mathbb{C}, \ |z| = 1\}$ and $SU(N)$ of $U(N)$, which are the images of the factor groups. Being a direct sum of two nonisomorphic simple Lie algebras, $\mathfrak{u}(N)$ thus admits, for any $N \geq 2$, a 2-dimensional family of Ad-invariant symmetric bilinear forms. See also §6.8.

(ii) As a structure group for the electroweak model, $U(2)$ has the following advantages over $S^1 \times SU(2)$:

   a) Its definition is simpler, and geometrically more natural;

   b) It has a faithful 2-dimensional representation;

   c) One could try to introduce a model based on $S^1 \times SU(2)$, instead of $U(2)$, by assuming that the electroweak interaction bundle is explicitly given as a tensor product $\iota = \chi\tau$ of complex vector bundles χ, τ of fibre dimensions 1 and 2, with structure groups $U(1)$ and $SU(2)$. Then $\Lambda^2\tau = 1$ (as in (viii) of §7.3), which leads, by (6.24) and (5.5), to $\lambda = \Lambda^2\iota = \chi^2\Lambda^2\tau = \chi^2$. In other words, χ then would be an *unphysical* electromagnetism bundle representing half of the electron charge.

(iii) An interaction may be of a *long* or *short range* depending on how fast its strength decreases with increasing distance. (A more rigorous definition involves asymptotic decay rates, as illustrated by Remark 6.9.3.) Gravitational, electromagnetic and strong (interquark) interactions are long-range, while the weak interaction is short-range. Residual interactions (§1.4), such as nuclear or interatomic forces, are short-range;

however, the weak interaction does not seem to be the residual effect of something else.

(iv) In most unified descriptions each particle species stands both for itself and, implicitly, for its antiparticle. The antiparticles could be explicitly brought into the picture by replacing each bundle η (e.g., α_{ew} in (6.21)) with $\eta + \bar{\eta}$. This, however, may lead to redundancy (if η has real summands), as well as ambiguities: Is $\sigma_L + \sigma_R$ a neutrino and an antineutrino, or an electron stripped of its electric charge?

(v) The generic-lepton bundle for the electroweak model that simultaneously describes k generations of leptons is the direct sum $k\sigma = \sigma + \ldots + \sigma$ of k copies of a Dirac spinor bundle σ over $(\mathcal{M}, g)$, where each summand is explicitly identified with a generation (*generation-labeled*). Instead of (6.21) and (6.26) we then have $\alpha_{\mathrm{ew},k} = k\alpha_{\mathrm{ew}} = k(\sigma_L + \lambda\sigma)$, which leads to k generation-labeled, neutrino-massive lepton pairs. The interaction bundle ι, the choice of p_0, q_0, ϕ, and the resulting decomposition (6.30) of $\mathcal{C}(\iota)$ leading to the interaction carriers $\gamma, \mathrm{W}^-, \mathrm{W}^+$ and Z^0, remain unchanged.

(vi) Due to its weakness, the weak interaction does not lead to bound states of physical interest, while those formed by the electromagnetic interaction have been discussed in §6.1. This is one of the reasons why we do not study bound states in the electroweak model.

(vii) Discussing the electroweak model, we have to assume that the space-time $(\mathcal{M}, g)$ is orientable, so that σ_L and σ_R can be globally separated. However, $\mathcal{M}$ essentially need not be *oriented* in the electroweak theory even if formula (6.21) seems to indicate that, by assigning σ_L a role considerably different from that of σ_R. The orientation of $\mathcal{M}$ implicit in (6.21) can namely be reversed if we decide to describe antileptons instead of leptons, i.e., replace $(e, \nu_e), \sigma, \sigma_L, \sigma_R, \iota$ and α_{ew} by $(e^+, \bar{\nu}_e), \bar{\sigma}, \bar{\sigma}_L = \sigma_R, \bar{\sigma}_R = \sigma_L, \bar{\iota}$ and $\bar{\alpha}_{\mathrm{ew}}$, respectively.

(viii) The discussion in §6.5 would be exactly the same without requiring the positive function $|\phi|$ on $\mathcal{M}$ to be constant. However, constancy of $|\phi|$ can be justified using arguments related to the dynamics of the electroweak model (§11.5). Furthermore, choosing ϕ such that $|\phi|$ is nonconstant, one would enrich the geometry in question far beyond the required structure-group reduction.

(ix) Given a subbundle ζ of a vector bundle η and a projection morphism $\mathrm{pr}: \eta \to \zeta$, we can *restrict* any connection ∇ in η to a connection $\nabla^{[\zeta]}$ in ζ with $\nabla_v^{[\zeta]}\psi = \mathrm{pr}(\nabla_v\psi)$ for sections ψ of ζ and tangent vectors v. If ∇ is compatible with a fibre metric $\langle,\rangle$ in η and pr is the orthogonal projection, then $\nabla^{[\zeta]}$ is compatible with the restriction of $\langle,\rangle$ to ζ.

(x) The (orthogonal) projection $\mathcal{C}(\iota) \to \mathcal{C}_\phi(\iota)$ in (6.27) acts on sections ∇ of $\mathcal{C}(\iota)$ by $\nabla \mapsto \nabla^{\{\phi\}} \oplus \nabla^{[\lambda]}$, where $\nabla^{\{\phi\}}$ is the flat connection in $\mathbb{C}\phi$ making ϕ parallel, while $\nabla^{[\lambda]}$ is obtained as in (ix) using the orthogonal projection pr of ι onto $\lambda = \phi^\perp$, and so their direct sum is a connection in $\iota = \mathbb{C}\phi \oplus \lambda$.

References: Ryder 1985, 1986; Leite Lopes 1981; Huang 1982; Sudbery 1986.

6.7 The Standard Model of Particles and Interactions

The *standard model* of elementary particles comprises those parts of particle theory which physicists regard as basically correct in view of far-reaching agreement between their predictions and experimental data. It does not include physical ideas whose status in this respect is a subject of controversy (such as grand unified theories, cf. §7.1), or cannot be decided using currently available experimental methods (e.g., supersymmetry or string theory).

The structure of the standard model allows to naturally distinguish its two levels (or "layers"), one consisting of the *field quantization* procedure, the other encompassing the *classical* (i.e., non-quantum) aspects of the theory. It is the field quantization methods, not discussed here, that lead to predictions consistent with experimental evidence.

The classical layer of the standard model includes its basic geometry and dynamics. Specifically, it consists of the quark model (§6.2), the electroweak model (§6.5) along with its dynamics (§11.5), and an extension of the electroweak theory to quarks (§9.4). In particular, according to the standard model, the only existing particles are the interaction carriers listed in §1.5, leptons (§1.14), quarks (§6.2), and bound states of these. However, the standard model does not explain why there are exactly three known lepton or quark generations (6.20), (6.1), nor does it preclude the existence of further, heavier quarks or leptons, the discovery of which might possibly be expected when sufficiently high accelerator energies become available (cf. Remark 6.9.5), even though, according to some experimental data, such discoveries do not seem likely.

A presentation of the standard model on its classical level is the main objective of this book. A few quantization-based arguments, scattered throughout the text (§§4.8, 4.9 as well as §§1.14, 1.16, 2.6, 2.7) do not correspond to the *field* or *second* quantization, but rather refer to the *first* quantization procedure (§2.5), used to provide semiclassical derivations for some ingredients of the classical theory.

The classical level of the standard model, taken on the face value, does not adequately describe the particle world known from experiments, as it is based on a macroscopic approximation (the classical limit). Its relation to genuine physical phenomena should therefore be treated with extreme caution. However, the classical model provides a natural foundation for the quantum theory, which can be built upon it using field quantization techniques. Moreover, despite its failure to provide reasonably accurate quantitative predictions, the classical part correctly accounts for many qualitative aspects of the microworld. The latter include quantum numbers such as spin, parity, or the electric charge, whose values for the particles involved, rendered by the geometric models in §§6.1, 6.2, 6.5, are in perfect agreement with reality.

6.8 Invariant Inner Products and Coupling Constants*

Let δ be a general interaction bundle of fibre dimension N, with a structure group $G \subset \mathrm{GL}(N,\mathbb{C})$, over the given spacetime $(\mathcal{M}, g)$, as in §5.4. It is often necessary to select a fibre metric $\langle,\rangle$ in the Lie algebra bundle $\mathfrak{g}(\delta)$ (§5.2), which is *natural* in the sense that it comes from an Ad-invariant inner product on the matrix Lie algebra $\mathfrak{g}$ corresponding to G, i.e., from a biinvariant metric on G. If $G = \mathrm{U}(N)$ or $G = \mathrm{SU}(N)$, $\langle,\rangle$ is given by

$$\langle a, b \rangle = -p_0[N\,\mathrm{Trace}\,(ab) - \mathrm{Trace}\,a \cdot \mathrm{Trace}\,b] - q_0\,\mathrm{Trace}\,a \cdot \mathrm{Trace}\,b \quad (6.32)$$

or, respectively,

$$\langle a, b \rangle = -Np_0\,\mathrm{Trace}\,(ab) \quad (6.33)$$

for $a, b \in \mathfrak{g}(\delta_x)$, $x \in \mathcal{M}$ (see (i) of §6.6), with real constants p_0, q_0 which are positive, since we prefer $\langle,\rangle$ to be positive definite. The positive numbers

$$C_0 = \frac{1}{\sqrt{p_0}}, \quad \tilde{C}_0 = \frac{1}{\sqrt{q_0}} \quad (6.34)$$

then are called the *coupling constants* corresponding to $\langle,\rangle$. Obviously, we are only interested in those of the numbers (6.34) which are uniquely determined by $\langle,\rangle$, as is the case for C_0 when $N \geq 2$, and for $\tilde{C}_0$ if $G = \mathrm{U}(N)$.

Besides the algebraic definition (6.32) – (6.34), C_0 and $\tilde{C}_0$ have simple geometric characterizations in terms of diameters of normal subgroups of $\mathrm{U}(N)$ or $\mathrm{SU}(N)$ relative to the corresponding biinvariant metric. (See Remark 6.8.2 below.) The reason why we cannot eliminate the freedom in choosing C_0 and $\tilde{C}_0$ by prescribing these diameters is that the inner products involved are usually valued in some physical quantities rather than pure (dimensionless) numbers, and so any such normalization would amount to arbitrarily selecting a measurement unit.

We had to choose a natural fibre metric $\langle,\rangle$ in $\mathfrak{g}(\delta)$ when we discussed the electroweak model, with $\delta = \iota$, $G = \mathrm{U}(2)$ in §6.5 (and we will also need one for the SU(5) theory, with $\delta = \beta$, $G = \mathrm{SU}(5)$, in §7.1). In either case, the fibre metric induced by $\langle,\rangle$ in the affine bundle $\mathcal{C}(\delta)$ is used to obtain an orthogonal direct-sum decomposition of the latter. Such decompositions obviously remain unchanged if $\langle,\rangle$ is multiplied by a positive factor, which seems to make the choice of $\langle,\rangle$ for simple structure groups G (such as $\mathrm{SU}(N)$) totally irrelevant. However, the example of electromagetism, discussed next, indicates that this is not necessarily the case. In other words, one can give a natural physical meaning not only to the ratio of the coupling constants $C_0, \tilde{C}_0$, but also to the constants themselves.

Specifically, any Hermitian connection ∇ in the electromagnetism bundle λ (§4.10, §6.1) is represented, for a fixed local unit section ξ of λ, by two mutually proportional local 1-forms Γ, A, characterized by (4.22), (4.20) and related by (4.21), which may be thought of as two possible descriptions of ∇, one *geometric* (the connection form Γ), the other *physical* (the local potential A, obtained by taking an antiderivative of the directly observable electromagnetic field F, cf. §4.6). Both Γ and iA are local sections of the translation-space bundle $\mathfrak{u}(\lambda)T^*$ of $\mathcal{C}(\lambda)$, since Γ is the difference between two local sections of $\mathcal{C}(\lambda)$ (specifically, ∇ and the flat connection making ξ parallel). On the other hand, $\mathfrak{u}(\lambda) = \mathcal{M} \times i\mathbb{R}$ is canonically trivialized, so that $\mathfrak{u}(\lambda)T^*$ carries a natural pseudo-Riemannian fibre metric $\langle , \rangle_g$ which is the g-induced inner product of 1-forms on $(\mathcal{M}, g)$. However, the physical character of A suggests that a *different* inner product $\langle , \rangle$ should be applied to Γ, namely

$$\langle \Gamma, \Gamma \rangle = \langle A, A \rangle_g = \frac{\hbar^2}{q^2} \langle \Gamma, \Gamma \rangle_g \qquad (6.35)$$

(cf. (4.21)). In this way, we arrive at the fibre metric in $\mathcal{C}(\lambda)$, i.e., in $\mathfrak{u}(\lambda)T^*$, induced by g along with the fibre metric in $\mathfrak{u}(\lambda)$ given by (6.32) with $q_0 = \hbar^2 q^{-2}$, while p_0 is not defined. (Note that what we did is more than a mere change of scale, since the components of Γ and A represent different physical units.) By (6.34), the corresponding coupling constant is

$$\tilde{C}_0 = \frac{|q|}{\hbar}. \qquad (6.36)$$

Thus, up to the universal constant factor $1/\hbar$, $\tilde{C}_0$ for electromagnetism equals the absolute value of the elementary (electron) charge q. Extending this to other interactions, we can say that *the coupling constants corresponding to the natural metric one selects in $\mathfrak{g}(\delta)$ represent the basic strengths of the interaction described by δ.* (Here "basic" means "not modified", e.g., by multiple-charge formation as in §4.13.) The geometry-to-physics-ratio interpretation (4.21) of the coupling constant (6.36) makes sense for other interactions as well; see Remark 6.8.3.

Remark 6.8.1. The coupling constants $C_0 = C_{\text{ew}}$, $\tilde{C}_0 = \tilde{C}_{\text{ew}}$ of the electroweak model are given by (6.34) with p_0, q_0 as in (6.8). The corresponding fibre metric in $\mathcal{C}(\iota)$ then induces one given by (6.29) in $\mathcal{C}(\lambda)$, λ being the resulting electromagnetism bundle. Thus, by (6.36), the (conventionally negative) electron charge q satisfies $q = -\hbar(p_0 + q_0)^{-1/2}$. Consequently, (6.34) and (6.10) imply that

$$q = -\hbar C_{\text{ew}} \sin\theta = -\hbar p_0^{-1/2} \sin\theta, \qquad (6.37)$$

where θ is the Weinberg angle, with $\tan \theta = \tilde{C}_{\text{ew}}/C_{\text{ew}}$.

A similar discussion of the coupling constants in the SU(5) grand unified theory can be found in (xiv) of §7.3.

Remark 6.8.2.   The coupling constants (6.34) can also be characterized by the relations

$$C_0 \operatorname{diam} \mathrm{SU}(N) = N\pi, \quad N \geq 2, \quad \tilde{C}_0 \operatorname{diam} \mathrm{S}^1 = N\pi,$$

involving the unique nontrivial, proper, normal connected subgroups SU(N), S^1 of G, if $G = \mathrm{U}(N)$ (cf. (i) of §6.6), or the group SU(N) itself, if $G = \mathrm{SU}(N)$. Their diameters are measured using the biinvariant metric on G corresponding to (6.32) or (6.33). In fact, this is easily established by connecting group elements to Id with geodesics (one-parameter subgroups) lying in maximal tori, described in turn as groups of linear isometries of $\mathbb{C}^N$ diagonalized by fixed orthonormal bases.

Relation (6.19) now follows easily from (6.10) and (6.34).

Remark 6.8.3.   Given a general interaction bundle δ with a simple structure group G such as SU(N), along with a local system $\xi_1, \ldots, \xi_N$ of trivializing sections of δ, compatible with its G-structure, and any compatible connection ∇ in δ, one can generalize (4.21), using (6.36), so as to define the corresponding $N \times N$ matrix of *potentials* $[A_\alpha^\beta]$ for ∇ and the ξ_α by

$$\Gamma_\alpha^\beta = \pm \mathrm{i} C_0 A_\alpha^\beta, \tag{6.38}$$

where Γ_α^β are the connection forms with $\nabla_v \xi_\alpha = \Gamma_\alpha^\beta(v)\xi_\beta$ for tangent vectors v and C_0 is an appropriate coupling constant, while the choice of the sign is a matter of convention. For reducible structure groups G, several coupling constants may be needed, corresponding to the components of the $\mathfrak{g}$-valued 1-form Γ (with the entries Γ_α^β) relative to the direct-sum decomposition of $\mathfrak{g}$. Thus, if $G = \mathrm{U}(N)$, we have $\Gamma_\alpha^\beta = \pm \mathrm{i}[C_0(A_\alpha^\beta - \frac{1}{N}A_\gamma^\gamma\delta_\alpha^\beta) + \frac{1}{N}\tilde{C}_0 A_\gamma^\gamma \delta_\alpha^\beta]$. Regarding the potentials A_α^β as some kind of "physical" quantities, as opposed to the "geometric" connection components Γ_α^β, we can again interpret the coupling constants as geometry-to-physics conversion factors. The analogy with the case of electromagnetism is further stressed by the equality $\langle \Gamma, \Gamma \rangle_{p_0, q_0} = \langle A, A \rangle_{1,1}$, cf. (6.35), where $\langle , \rangle_{p_0, q_0}$ is defined (also for $p_0 = q_0 = 1$) by (6.32) or (6.33), while p_0, q_0 are related to $C_0, \tilde{C}_0$ by (6.34), and q_0 is to be dropped if $G = \mathrm{SU}(N)$.

References: Ryder 1986; Bogolyubov and Shirkov 1983.

6.9 Massiveness Versus Masslessness*

A somewhat surprising consequence of the electroweak model (§6.5) is that, despite the seemingly fundamental difference between nonzero and zero masses, massive particles may coexist with massless ones in the same "multiplet" in the sense of being represented, before symmetry breaking, by a common generic particle. This applies both to matter particles, (in this case, leptons) with masses satisfying (7.1), and to the electroweak interaction carriers γ, $W^{\pm}$, Z^0, for which $m_\gamma = 0 < m_W < m_Z$ (Remark 6.9.1). We will be able to partly "explain" this phenomenon (i.e., predict it on the basis of other, plausible assumptions) only when discussing dynamics of the electroweak model in §11.5. At present, we must content ourselves with a more formal view of the massive/massless distinction, focusing on the question *how*, rather than *why*, some particle species come to have zero mass.

In nature, massless particles apparently consist of very few species, namely, the neutrinos ν_e, ν_μ, ν_τ, the corresponding antineutrinos $\overline{\nu}_e, \overline{\nu}_\mu, \overline{\nu}_\tau$, the photon γ, and 8 kinds of gluons (§2.14). Their models, listed below, differ considerably from those used to describe massive particles. Specifically, the former cannot be obtained from the latter just by setting the mass parameter m equal to 0. This is consistent with the idea, based on common sense and our macroscopic experience, that positive masses consititute the norm, while masslessness is a peculiar aberration. In other words, a particle is to be *presumed massive until proven massless*, the only kind of admissible evidence being the impossibility of plausibly inserting a *mass term* with $m > 0$ into a description of the particle.

Among matter particles, the only massless ones are the (anti)neutrinos, represented by Weyl spinor bundles σ_L, σ_R with the Weyl equation $\mathcal{D}\psi = 0$ imposed on their sections ψ. (See §1.16 and §6.3.) Masslessness of the neutrinos can also be deduced from Weyl's equation (at least, over spacetimes with Scal $= 0$), as it leads, by (1.8), to the Klein-Gordon equation (1.3) with $m = 0$. Moreover, the very use of the bundle σ_L or σ_R reasonably justifies the conclusion that the mass m of each neutrino is zero. In fact, with $m > 0$, we would have to replace Weyl's equation by the Dirac equation (1.7) (see §3.9) which, however, has no non-trivial solution ψ in σ_L or σ_R, as $\mathcal{D}$ sends sections of σ_L onto those of σ_R and vice versa. (Solutions ψ in σ_L would exist if we identified $\sigma_L = \overline{\sigma_R}$ and σ_R as *real* bundles, but then we would end up with the relatively unnatural task of solving an *antilinear* eigenvalue problem.)

The remaining massless particles, the photon and the gluons, are interaction carriers for the electromagnetic and strong (interquark) interaction, respectively. These interactions are described by the Yang-Mills connection scheme without spontaneous symmetry breaking (§6.1, §6.2), which formally implies masslessness of their carriers. In fact, in contrast with (1.3) or (1.7), neither the (usually nonlinear) Yang-Mills equation (5.7), nor its linearization at any fixed solution ∇ contains a naturally identifiable mass term (although either of them displays the corresponding "d'Alembertian term"). For details, see Remark 6.9.4.

On the other hand, if spontaneous symmetry breaking (§5.8, §6.5) makes a part of the connection components appear as matter fields, i.e., sections of the vector bundles listed in §§1.16, 1.17 (other than the neutrino models), we expect the corresponding interaction carriers to be massive, in view of the preceding discussion.

Remark 6.9.1. Some ideas in cosmology are based on the presumption that neutrino masses are positive, although very small. Available experimental data only provide upper bounds for neutrino masses, such as $m_{\nu_e} < 0.000046$ MeV/c^2 for the electronic neutrino. Compared to the (approximate) masses $m_e = 0.5110034$, $m_p = 938.2796$, $m_W = 80800$ and $m_Z = 92900$ of the electron, proton, W boson and Z boson (all in MeV/c^2), this makes masslessness of neutrinos appear quite plausible. However, should neutrinos turn out to be massive as a result of lower bounds on their masses that some future experiments might provide, the neutrino models currently in use would obviously have to be changed. As the preceding discussion indicates, the required modifications would consist in replacing a Weyl spinor bundle σ_L or σ_R with the Weyl equation (1.6), by the Dirac spinor bundle $\sigma = \sigma_L + \sigma_R$, or its conjugate $\bar{\sigma}$, with Dirac's equation (1.7), m being the now-positive mass of the neutrino. The orientation of space or spacetime distinguished by each individual (anti)neutrino species, i.e., its (left or right) handedness, originally reflected by the choice between σ_L and σ_R, now would be expressed by σ being "chiral" in the sense that its summands σ_L, σ_R are *ordered*, in contrast with the case of (6.4).

Remark 6.9.2. If the neutrinos are found to be massive, and hence described as in Remark 6.9.1, the electroweak model (§6.5) can easily be modified so as to become not only consistent with the massive-neutrino description, but also simpler and more natural in its own right. The modification in question consists in replacing the interacting-lepton bundle $\alpha_{ew} = \iota\sigma_L + (\Lambda^2\iota)\sigma_R$ by the "more fundamental" expression $\alpha = \iota\sigma$ (cf. §5.4). Instead of (6.27) one then obtains $\alpha = \sigma + \lambda\sigma$, which accounts for the massive neutrino and the electron, both with with their correct electric charges. The electroweak interaction bundle ι, the generic-lepton bundle σ, the "symmetry-breaking device" ϕ_0 and the resulting description (6.31) of the interaction carriers, remain otherwise unchanged, except that the σ_L and σ_R summands of σ in $\alpha = \sigma + \lambda\sigma$ are to be *ordered*, while those for $\lambda\sigma$ are not. The last requirement can also be thought of as "attaching" an orientation of the spacetime to the summand $1 = \mathbb{C}\phi$ in the decomposition $\iota = 1 + \lambda$ (i.e., to the section ϕ of ι selected in the process of symmetry breaking).

Remark 6.9.3.   An interaction is short-range (long-range) if and only if its carrier particles are massive (massless). This fact, pointed out by Yukawa in 1935, can be illustrated as follows. Let the interaction carrier of mass $m \geq 0$ be represented by a natural bundle η over the Minkowski spacetime M. For simplicity, we use an identification $\eta = M \times W$ and regard sections ψ of η as functions on M valued in a vector space W, and satisfying the Klein-Gordon equation (1.3). Choosing an observer in M (§4.2), we obtain $M = \mathbb{R} \times \mathbf{V}$, where $\mathbf{V}$ is the observer's Euclidean 3-space.

Assume now that the state ψ of the carrier particle is such that it stays at the origin $\mathbf{0} \in \mathbf{V}$ without changing with time or distinguishing any direction in space, i.e., $\psi(x) = f(r)$ for some function f and $x \in M$, where $x \sim (ct, \mathbf{x})$ (notation of §4.2) with $r = |\mathbf{x}|$. Then (1.3) is equivalent to the equation $d^2 f/dr^2 + 2r^{-1} df/dr - m^2 c^2 \hbar^{-2} f(r) = 0$, the solutions of which have the form

$$f(r) = r^{-1}[\psi_0 \exp(-mcr/\hbar) + \phi_0 \exp(mcr/\hbar)], \quad \text{if} \quad m > 0,$$

or

$$f(r) = \psi_0 r^{-1} + \phi_0, \quad \text{if} \quad m = 0,$$

with constants $\psi_0, \phi_0 \in W$. Moreover, $\phi_0 = 0$, since we expect ψ, as a function on $\mathbf{V}$, to vanish at infinity or be in some L^p away from the origin, in order for the energy of the particle to be finite. (See the medium-disturbance interpretation in (a) of §1.2.) Consequently,

$$\psi(x) = \frac{\psi_0}{|\mathbf{x}|} \exp\left(-\frac{mc}{\hbar}|\mathbf{x}|\right)$$

for all $m \geq 0$. (We ignore the singularity of ψ at $\mathbf{x} = \mathbf{0}$.) Thus, if $m > 0$, ψ and its partial derivatives of all orders decay exponentially as $|\mathbf{x}| \to \infty$, in contrast with the $|\mathbf{x}|^{-1}$ asymptotic decay rate when $m = 0$. Since the range of the interaction is obviously related to the distance at which states of its carriers remain significantly nonzero, our assertion follows.

Remark 6.9.4.   Interaction carriers represented by connections in an interaction bundle (§5.4), without spontaneously broken symmetry (§5.8), are massless, which can be justified as follows. Using a local trivialization of a given vector bundle η over a spacetime $(\mathcal{M}, g)$ and a local coordinate system in $\mathcal{M}$, one can easily rewrite the Yang-Mills equation div $R^\nabla = 0$, for connections ∇ in η, in the form

$$\partial \partial \Gamma + \Gamma \partial \Gamma + \Gamma \Gamma \Gamma = 0, \tag{6.39}$$

where Γ stands for the components of the connection form of ∇, while $\partial \Gamma$ and $\partial \partial \Gamma$ are suitable expressions linear in the first (second) order partial derivatives of the Γ, and multiplications represent specific multilinear maps involving appropriate numbers of arguments. Linearizing (6.39) at a fixed ∇

(or Γ), i.e., differentiating it with respect to a parameter τ that Γ depends on, we obtain $\partial\partial\dot{\Gamma}+\Gamma\partial\dot{\Gamma}+\dot{\Gamma}(\partial\Gamma+\Gamma\Gamma)+\Gamma\dot{\Gamma}\Gamma+\Gamma\Gamma\dot{\Gamma}=0$ with $\dot{\Gamma}=d\Gamma/d\tau$. Neither of these equations contains a nonzero mass term such as in (1.3) or (1.7), i.e., a constant multiple of the unknown Γ or $\dot{\Gamma}$.

Remark 6.9.5. The proportionality relation betwen the mass m of any pointlike object and its rest energy mc^2 (§4.3) indicates that mass and energy are "equivalent" and can be converted into each other. In fact, heavy particles are often created in collisions involving much lighter objects with sufficiently high kinetic energies. However, although the total relativisitic energy E is a universally conserved additive quantity, we cannot expect in general that the whole sum $E = \sum_\alpha E_\alpha$ of the constituent-particle energies will turn into the mass of the particle produced in the collision, as some part of E may have to become the kinetic energy. Specifically, the mass m of a particle which is the unique product of a collision is given by

$$m = \frac{1}{c^2}\sum_\alpha E_\alpha, \tag{6.40}$$

where E_α are the energies of the colliding particles measured with respect to the whole system's (moving) center of mass. (See Remark 6.9.6.) Therefore, as m increases, it becomes increasingly difficult to discover new particles of mass m, which is done by producing them in collisions, since one then needs accelerators capable of generating colliding systems of particles with sufficiently high center-of-mass energies.

Remark 6.9.6.   For the purpose of a relativistic discussion of collisions involving several point-like particles of masses m_α, it is useful to introduce the 4-momentum $p = \sum_\alpha p_\alpha$ of the whole system, where p_α are the 4-momenta (§4.3) of the constituents at the collision event (which is a point where their world lines intersect). Since p is again a future-pointing timelike vector, it is natural to use formula (1.10), i.e., $|p| = mc$, to define the *combined mass* m of the system at the collision event. In the case where the collision produces just a single particle, p and m will be the 4-momentum and mass of that particle, which amounts to the 4-*momentum conservation law.* The easily established "reversed" Schwarz inequality $|\sum_\alpha p_\alpha| \geq \sum_\alpha |p_\alpha|$ for future-pointing timelike vectors p_α implies that

$$m \geq \sum_\alpha m_\alpha,$$

with some excess mass created out of the collision energy. (Both inequalities are actually strict unless all p_α are collinear, i.e., unless the particles are mutually at rest.) Choosing an observer which is at rest relative to the system's center of mass, in the sense that $\mathbf{p} = \sum_\alpha \mathbf{p}_\alpha = \mathbf{0}$ in (4.7) (that is, an observer (v_0, y_0) with $v_0 = p/|p|$, cf. §4.2), we obtain, from (1.10), $mc^2 = E = \sum_\alpha E_\alpha$. This implies (6.40).

Our discussion remains valid if the collision involves massless particles as well, so that, for some (or all) of the α, $m_\alpha = 0$ and p_α is a nonzero future-pointing *null* vector (Remark 4.3.2). In fact, p will then still be timelike, as we may obviously assume that the p_α are not all collinear.

Reference: Felsager 1983.

7 Grand Unified Theories

Attempts of going beyond the standard model to achieve a unification of all particle interactions lead to constructions called *grand unifications*, or grand unified theories. Two such theories are discussed in this chapter.

7.1 The SU(5) Grand Unification

The unification of the weak and electromagnetic interactions that was so successfully carried out in the electroweak model (§6.5), naturally suggests a further step in the form of a similar theory unifying *all* microworld interactions, i.e., the electroweak and strong ones. Models of this kind are known as *grand unified theories* and many of them, involving more than a dozen of different structure groups, have been proposed in the physical literature since early 1970s. (See (i) of §7.3.)

All existing grand unified theories lead to serious problems, illustrated by (v) of §7.3, which make their adequacy as models of physical reality at least questionable. This is why none of the grand unifications developed so far has been included in the standard model of elementary particles (§6.7).

The quark and lepton generations listed in (6.1) and (6.20) can be regarded as ordered by increasing "generation masses", in view of (6.2) and the lepton-mass relations

$$m_{\nu_e} = m_{\nu_\mu} = m_{\nu_\tau} = 0 < m_e < m_\mu < m_\tau. \tag{7.1}$$

Therefore, it is natural to combine them into three (*basic*) *fermion generations*

$$(e, \nu_e, u, d), \quad (\mu, \nu_\mu, c, s), \quad (\tau, \nu_\tau, t, b), \tag{7.2}$$

each of which represents the corresponding antiparticles as well, cf. (iv) of §6.6. The electric-charge pattern for each generation is $(-1, 0, \frac{2}{3}, -\frac{1}{3})$ in proton-charge units. The particles listed in (7.2), along with their antiparticles and the resulting bound states, give rise to all known species of matter particles.

Most grand unifications deal with each fermion generation in (7.2) separately. The corresponding model for several generations then can be obtained as in (v) of §6.6. This is also the case for the SU(5) *grand unified theory*, formulated by Georgi and Glashow in 1974, which is the simplest of all grand unifications. We now proceed to discuss its geometry, choosing the fermion generation, for definiteness, to be (e, ν_e, u, d). Another grand unification, based on the group Spin(10), is outlined in §7.2.

As in the electroweak case, we begin with a free-particle bundle, here referred to as a *generic-fermion bundle*, which is a fixed Dirac spinor bundle $\sigma = \sigma_L + \sigma_R$ over the given orientable spacetime $(\mathcal{M}, g)$. A suitable version of the Yang-Mills scheme of §5.4, along with spontaneous symmetry breaking, then leads as in §6.5 to a description of the resulting interacting matter particles and interaction carriers.

Specifically, the interaction bundle β for the SU(5) model is a complex vector bundle of fibre dimension *five*, with a fixed Hermitian fibre metric $\langle,\rangle$ and a compatible volume form Θ, over $\mathcal{M}$, so that its structure group is SU(5) (cf. (iii) of §5.2). The interacting-particle (*interacting-fermion*) bundle then is defined, in complete analogy with (6.21), by the formula

$$\alpha_{gu} = \beta\sigma_L + (\Lambda^2\beta)\sigma_R. \tag{7.3}$$

The kind of spontaneous symmetry breaking involved in the SU(5) theory can be conveniently split into three consecutive steps. The *first step* consists in selecting a complex vector subbundle ι of β, of fibre dimension 2. Endowed with the Hermitian fibre metric induced by $\langle,\rangle$, ι is referred to, and physically interpreted as, the *electroweak-interaction bundle* (§6.5). We have $\beta = \iota + \iota^\perp$, so that $\Lambda^2\beta = \Lambda^2\iota + \Lambda^2\iota^\perp + \iota\iota^\perp$ (cf. (vii) of §7.3) and hence, by (7.3),

$$\alpha_{gu} = \alpha_{ew} + \alpha' \tag{7.4}$$

with $\alpha_{ew} = \iota\sigma_L + (\Lambda^2\iota)\sigma_R$, as in (6.21), and

$$\alpha' = \iota^\perp\sigma_L + (\Lambda^2\iota^\perp)\sigma_R + \iota\iota^\perp\sigma_R. \tag{7.5}$$

Inspired by (6.24), we can provisionally define the *electromagnetism bundle* representing the electron charge to be the Hermitian line bundle $\lambda = \Lambda^2\iota$, later to be identified with a subbundle of ι.

The *second step* of symmetry breaking amounts to choosing a *cubic root* for λ, i.e., a complex Hermitian line bundle $\lambda^{1/3}$ along with a fixed isometric bundle isomorphism $(\lambda^{1/3})^3 = \lambda$. (About the existence and uniqueness questions, see (xii) of §7.3.) Choosing $\lambda^{1/3}$ is obviously necessary to handle the fractional electric charges of the quarks (§6.2).

According to (7.19) we have $1 = \Lambda^5\beta = (\Lambda^2\iota)\Lambda^3\iota^\perp = \lambda\Lambda^3\iota^\perp$ (notation of (5.1)), the trivialization of $\Lambda^5\beta$ being established, with the aid of Θ, as in (viii) of §7.3. Therefore, $\lambda = \overline{\Lambda^3\iota^\perp} = \Lambda^3\overline{\iota^\perp}$ and so, setting

$$\rho = \lambda^{-1/3}\overline{\iota^\perp}, \tag{7.6}$$

where the notation of (4.29) has been extended to fractional powers of λ with exponents of denominator 3, we have from (5.5) the natural isomorphism

$$\Lambda^3 \rho = 1. \tag{7.7}$$

Thus, the complex vector bundle ρ of fibre dimension 3 carries, in addition to the Hermitian fibre metric induced from $\langle, \rangle$, also a compatible volume form determined by (7.7), so that its structure group is SU(3) (cf. (viii) of §7.3). It is therefore plausible to interpret ρ as the *strong interaction bundle* (§6.2).

The surjective morphism $\rho \Lambda^2 \rho \to \Lambda^3 \rho = 1$ of exterior multiplication establishes a natural isomorphism $\Lambda^2 \overline{\rho} = \overline{\Lambda^2 \rho} = \rho$ (cf. (vi) of §7.3). On the other hand, (7.6) implies

$$\iota^\perp = \lambda^{-1/3} \overline{\rho} \tag{7.8}$$

and hence, by (5.5), $\Lambda^2 \iota^\perp = \lambda^{-2/3} \rho$. Consequently, (7.5) can be rewritten as

$$\alpha' = \lambda^{-1/3} \overline{\rho} \sigma_{\rm L} + \lambda^{-2/3} \rho \sigma_{\rm R} + \lambda^{-1/3} \iota \overline{\rho} \sigma_{\rm R}. \tag{7.9}$$

The *third step* of symmetry breaking consists, exactly as in §6.5, in selecting a nonzero constant-length section ϕ of ι. From now on we will use (6.25) to identify $\lambda = \Lambda^2 \iota$ with the subbundle $\phi^\perp$ of ι. Thus, $\iota = \mathbb{C}\phi + \phi^\perp = 1 + \lambda$ and, by (7.4) and (7.9) along with (6.3), (6.4),

$$\alpha_{\rm gu} = \sigma_{\rm L} + \lambda\sigma + \lambda^{-1/3} \overline{\rho}\sigma + (\overline{\lambda^{-2/3}\rho\sigma_{\rm L}} + \lambda^{-2/3}\rho\sigma_{\rm R}). \tag{7.10}$$

In this way, the generic first-generation fermion, subject to the unified interaction and spontaneous symmetry breaking, becomes a generalization (§1.8) of four physical fermion species, corresponding to the four direct summands in (7.10). The first three of them are unmistakably the neutrino $\nu_{\rm e}$, electron e and d-antiquark $\overline{\rm d}$ (see, however, (xiii) of §7.3), represented by the free-particle bundles $\sigma_{\rm L}, \sigma$ along with the factors $\lambda, \lambda^{-1/3}, \overline{\rho}$ correctly accounting for their electric charges (Remark 4.13.1), and for whether and how they participate in the strong interaction (§6.2). As for the fourth summand, it is obtained from the correct model

$$\lambda^{-2/3}\rho\sigma = \lambda^{-2/3}\rho\sigma_{\rm L} + \lambda^{-2/3}\rho\sigma_{\rm R}$$

of the u quark subject to the electromagnetic and strong interaction (§§6.1, 6.2), by replacing the left-handed summand $\lambda^{-2/3}\rho\sigma_{\rm L}$ with its conjugate bundle. Thus, except for the u quark being "folded over", the SU(5) model provides a description of the fermion generation $({\rm e}, \nu_{\rm e}, {\rm u}, {\rm d})$ with all required properties. It turns out that, for dimensional reasons, each grand unified theory has to represent some particle species in a geometrically unsatisfactory manner, as for the u quark above. (See (iv) of §7.3.)

As in §6.5, the interaction carriers in the SU(5) model will be obtained from a direct sum decomposition of $\mathcal{C}(\beta)$, orthogonal with respect to a natural fibre metric $\langle, \rangle$ in $\mathcal{C}(\beta)$ (i.e., in its translation-space bundle $\mathfrak{su}(\beta)T^*$ with $T^* = T^*\mathcal{M}$). This time the choice of $\langle, \rangle$, that is, of an Ad-invariant inner product in the Lie algebra $\mathfrak{su}(5)$, is unique up to a factor, since the group SU(5) is simple (in contrast with U(2)). Specifically,

$$\langle a, b \rangle = -5r_0 \text{Trace}\,(ab) \tag{7.11}$$

with some real r_0 (assumed positive to ensure positive definiteness of $\langle , \rangle$) and all $a, b \in \mathfrak{su}(\beta_x)$, $x \in \mathcal{M}$. This uniquely determines the Weinberg angle θ for the electroweak model, now regarded as "embedded" in the $SU(5)$ theory. In fact, $\mathfrak{u}(\iota)$ may be identified with a subbundle of $\mathfrak{su}(\beta)$, with the inclusion morphism

$$\mathfrak{u}(\iota_x) \ni a \mapsto a \oplus (-\frac{1}{3}\,\text{Trace}\,a)\text{Id}_{\iota_x^{\perp}} \in \mathfrak{su}(\iota_x \oplus \iota_x^{\perp}) = \mathfrak{su}(\beta_x), \quad x \in \mathcal{M}\,(7.12)$$

characterized as taking the simplest traceless extension. Restricting (7.11) to $\mathfrak{u}(\iota) \subset \mathfrak{su}(\beta)$, we obtain (6.8) with

$$p_0 = \frac{5}{2}r_0\,, \qquad q_0 = \frac{25}{6}r_0 \tag{7.13}$$

and so, from (6.10),

$$\sin^2\theta = \frac{3}{8} = 0.375. \tag{7.14}$$

The Weinberg angle θ obtained from the SU(5) model therefore differs considerably from the experimental value (6.31). See also (iii) of §7.3.

The connections ∇ in β compatible with the SU(5)-structure are sections of the real affine bundle $\mathcal{C}(\beta)$ (§5.3). Those among them that leave ι (and hence $\iota^{\perp}$) invariant, in the sense that their parallel displacements do so, coincide with the sections of an affine subbundle $\mathcal{C}_\iota(\beta)$ of $\mathcal{C}(\beta)$ with the translation-space bundle $\mathfrak{su}_\iota(\beta)T^*$, where $\mathfrak{su}_\iota(\beta) \subset \mathfrak{su}(\beta)$ consists of all skew-adjoint traceless fibre endomorphisms in β which preserve ι. The orthogonal complement of $\mathfrak{su}_\iota(\beta)$ in $\mathfrak{su}(\beta)$ relative to the fixed fibre metric (7.11) we are using is obviously formed by those maps in $\mathfrak{su}(\beta)$ which interchange ι and $\iota^{\perp}$. By skew-adjointness, such maps are uniquely determined by their restrictions to $\iota_x^{\perp}$, $x \in \mathcal{M}$, valued in ι_x and otherwise arbitrary, which leads to the natural isomorphism $[\mathfrak{su}_\iota(\beta)]^{\perp} = \text{Hom}\,(\iota^{\perp}, \iota) = (\iota^{\perp})^*\iota = \overline{\iota^{\perp}}\iota = \lambda^{1/3}\rho\iota$ (cf. (7.8) and (vi) of §7.3). Consequently, we have the orthogonal direct-sum decomposition of affine bundles

$$\mathcal{C}(\beta) = \mathcal{C}_\iota(\beta) + \lambda^{1/3}\rho\iota T^*, \tag{7.15}$$

(which can also be obtained by describing the projection $\mathcal{C}(\beta) \to \mathcal{C}_\iota(\beta)$; see (xi) of §7.3). Another such decomposition is

$$\mathcal{C}_\iota(\beta) = \mathcal{C}(\rho) + \mathcal{C}(\iota), \tag{7.16}$$

where $\mathcal{C}(\rho)$, $\mathcal{C}(\iota)$ stand for the affine bundles whose sections are connections in ρ, ι compatible with the respective G-structure, G being SU(3) or U(2).

To obtain (7.16), note that, through obvious restrictions, a section ∇ of $\mathcal{C}_\iota(\beta)$ determines connections $\nabla^{[\iota]}$ in ι and $\nabla^{[\iota^\perp]}$ in $\iota^\perp$, which in turn naturally give rise to connections in $\overline{\iota^\perp}$, $\lambda = \Lambda^2\iota$, $\lambda^{-1/3}$ ((ix), (x) of §7.3) and, finally, to a connection $\nabla^{[\rho]}$ in $\rho = \lambda^{-1/3}\overline{\iota^\perp}$. The resulting assignment

$$\nabla \mapsto (\nabla^{[\rho]}, \nabla^{[\iota]}) \tag{7.17}$$

describes (7.16) on the level of sections. The morphism $\mathfrak{su}_\iota(\beta)T^* \to \mathfrak{su}(\rho)T^*$ $+\mathfrak{u}(\iota)T^*$ of the associated vector bundles, corresponding to (7.17), is the tensor product of Id_{T^*} with the morphism $\mathfrak{su}_\iota(\beta) \to \mathfrak{su}(\rho) + \mathfrak{u}(\iota)$ given by $a \mapsto (-\frac{1}{3}\mathrm{Trace}\, b \otimes \mathrm{Id}_{\overline{\iota^\perp}} + \mathrm{Id}_{\lambda^{-1/3}} \otimes c, \quad b)$ and (7.6), where b, c are the restrictions of a to ι and $\iota^\perp$, respectively. Therefore, (7.17) defines an affine bundle isomorphism, which establishes (7.16).

Combining (7.15) with (7.16) we obtain $\mathcal{C}(\beta) = \mathcal{C}(\rho) + \mathcal{C}(\iota) + \lambda^{1/3}\rho\iota T^*$. In view of the relations (6.23) and (6.30) obtained using ϕ selected above and the fibre metric in $\mathfrak{u}(\iota)$ induced by (7.11) in $\mathfrak{su}(\beta)$ via the embedding (7.12), we thus arrive at the orthogonal direct-sum decomposition

$$\mathcal{C}(\beta) = \mathcal{C}(\rho) + \mathcal{C}(\lambda) + \lambda T^* + T^* + \lambda^{4/3}\rho T^* + \lambda^{1/3}\rho T^*, \tag{7.18}$$

the summands of which represent, as in §1.8, the unified-interaction carriers predicted by the SU(5) theory. Specifically, these are

(i) The "eight kinds of" gluons, carriers of the strong (interquark) interaction, described by $\mathcal{C}(\rho)$ as in §6.2.

(ii) The photon γ, mediating the electromagnetic interaction and corresponding to $\mathcal{C}(\lambda)$ (§6.1).

(iii) The W and Z bosons $\mathrm{W}^-, \mathrm{W}^+, \mathrm{Z}^0$, carriers of the weak interaction, represented by λT^* and T^*, respectively, and formally behaving like matter fields (in contrast with the gluons and the photon). See the discussion following (6.30).

(iv) The X and Y mesons (bosons) $\mathrm{X}^{-4/3}, \mathrm{Y}^{-1/3}$ along with their antiparticles $\mathrm{X}^{+4/3}, \mathrm{Y}^{+1/3}$, described (cf. (iv) of §6.6) by the last two summands in (7.18), and hence interacting both electromagnetically (with fractional charges) and strongly. These bosons also resemble matter fields and so may be thought of as carriers of some additional short-range interaction. However, unlike the particle species in (i) – (iii), the X and Y mesons have not been detected experimentally. See also (v)c) of §7.3.

References: Mohapatra 1986; Leite Lopes 1981.

7.2 The Spin(10) Theory*

The Spin(10) *grand unification*, usually called the SO(10) theory, was proposed by Fritzsch and Minkowski in 1975. Its interaction bundle is a complex vector bundle ζ of fibre dimension 16, with a Spin(10)-structure over the spacetime $(\mathcal{M}, g)$. (Such a structure corresponds to a 16-dimensional faithful complex representation of Spin(10); see, e.g., [Derdzinski 1991a].) The whole fermion generation (e, ν_e, u, d) is represented by a generic-particle bundle which, in contrast with the SU(5) model, now is a fixed Weyl spinor bundle σ_L over $(\mathcal{M}, g)$ (§6.3). The interacting-particle bundle then is

$$\alpha = \zeta \sigma_L.$$

Simplicity of α, compared to α_{gu} given by (7.3) for the SU(5) theory, is an obvious advantage of the Spin(10) model. Its disadvantages include the prediction of many additional, still undiscovered interaction carriers, as $\dim \text{Spin}(10) = 45$ (while $\dim \text{SU}(5) = 25$).

Spontaneous symmetry breaking in the Spin(10) theory consists in first choosing a subbundle β of ζ with an SU(5) structure (as in §7.1) along with an identification $\zeta = 1 + \beta + \Lambda^2 \overline{\beta}$, and then proceeding with the three steps of SU(5) symmetry breaking described in §7.1. As a result, α becomes decomposed into summands representing e, ν_e, u and d, with a modified model of the neutrino ν_e (a chiral Dirac spinor bundle, as in Remark 6.9.1). The ensuing decomposition of $\mathcal{C}(\lambda)$ leads to interaction-carrier species, described in [Fritzsch and Minkowski 1975].

References: Fritzsch and Minkowski 1975; Mohapatra 1986; Leite Lopes 1981; Derdzinski 1991a.

7.3 Further Remarks

(i) The list of structure groups used for the existing grand unified theories includes SU(N) with $N = 5, 6, 7, 8, 9$, SO(N) or Spin(N) with $N = 10, 11, 12, 15, 16$, as well as E_6, E_7, E_8 and SU(8) × SU(8).

(ii) The very existence of the X and Y bosons leads, through arguments based on field quantization, to positive probabilities for interaction processes they mediate. Such processes include decays of quarks into leptons, thus implying nonconservation of the baryon number (see §§2.13, 2.14, 8.1) and instability of the proton. However, a decade of intensive searches brought no evidence of spontaneous proton decay, which indicates that the mean life of the proton, if finite at all, must be extremely long. In other words, if the X and Y exist, their masses are very large compared to all particles currently known, while the interaction they carry is very weak. Note that, as stated in §1.4, weakness of an interaction amounts to its infrequent occurrence, which in turn suggests that its carriers have to be rather "small", and hence, by Remark 6.9.3, quite massive.

(iii) The discrepancy between (6.31) and (7.14), i.e., between the experimental value of the Weinberg angle θ, and that predicted by the SU(5) theory, can be explained away using plausible theoretical arguments. Therefore, the incorrect value of θ alone does not invalidate the SU(5) model.

(iv) The bundles representing fermions other than neutrinos have the left and right-handed summands of equal fibre dimensions. Therefore, the fibre dimensions of these summands in the interacting-fermion bundle, such as (7.3), in any grand unified theory, should differ by at most twice the number of neutrino species predicted by the model. Since this requirement cannot be satisfied by interaction bundles β of fibre dimensions large enough to accomodate a unified theory, it is usually necessary for the resulting models of some matter particles to be "warped", as the u quark in the SU(5) case above, where a σ_R is used instead of a missing σ_L. Such features of the model, undeniably "ugly" from the viewpoint of geometry, could conceivably be eliminated by assuming the left and right-handed parts of the interacting-fermion bundle to be of *equal* dimensions. This might be done either by postulating positive neutrino masses (cf. Remark 6.9.2), or by allowing an even number of neutrino species, half of which would be represented by their antiparticles.

(v) The SU(5) model is plagued by several problems, typical of grand unified theories in general, such as

a) leading to predictions inconsistent with experimental evidence, which includes proton decay, some incorrect relations involving particle masses (obtained from the dynamics part of the theory, not discussed here), and the value of θ contradicting measurement data (see, however, (iii));

b) treating the fermion generations separately without accounting for their number;

c) not explaining the tremendous difference of masses between the X, Y and W, Z categories of interaction carriers (this is known as the *problem of gauge hierarchies*), cf. (ii) above.

(vi) For any complex vector bundle η carrying a distinguished Hermitian fibre metric $\langle,\rangle$, we use the natural identification $\eta^* = \bar{\eta}$ established by $\langle,\rangle$.

(vii) Given vector bundles η, ζ and an integer $k \geq 0$, we have the natural isomorphism $\Lambda^k(\eta + \zeta) = \sum_{j=0}^{k}(\Lambda^j \eta)\Lambda^{k-j}\zeta$ (notation of (5.1)), the embedding of each summand being given by $\alpha\beta \mapsto \alpha \wedge \beta$ along with the obvious inclusion $\Lambda^j W \subset \Lambda^j V$ whenever W is a subspace of a vector space V. In particular, if k, ℓ are the fibre dimensions of η and ζ, respectively, we obtain

$$\Lambda^{k+\ell}(\eta + \zeta) = (\Lambda^k \eta)\Lambda^\ell \zeta. \tag{7.19}$$

(viii) For a complex vector bundle η of fibre dimension N over $\mathcal{M}$, isomorphisms $F : \Lambda^N \eta \to 1 = \mathcal{M} \times \mathbb{C}$ are in a bijective correspondence with volume forms Θ in η ((ii) of §5.2), given by $\Theta(\psi_1, \ldots, \psi_N) = F(\psi_1 \wedge \ldots \wedge \psi_N)$. If, moreover, η is endowed with a Hermitian fibre metric $\langle , \rangle$, isomorphisms $F : \Lambda^N \eta \to 1$ which are *isometric* correspond in this way to volume forms Θ compatible with $\langle , \rangle$, i.e., to reductions of the structure group of η from $U(N)$ to $SU(N)$ (cf. (c) of §5.8).

(ix) For any complex vector bundle η with a G-structure, we have a natural isomorphism $\mathcal{C}(\eta) = \mathcal{C}(\bar\eta)$ of real affine bundles, established by the canonical antiisomorphism $\eta \to \bar\eta$ (§1.6).

(x) Let ζ be a Hermitian complex line bundle. For any integer $k \neq 0$, we have a natural isomorphism $\mathcal{C}(\zeta) = \mathcal{C}(\zeta^k)$ of real affine bundles, the homogeneous part of which, acting on the translation-space bundles $\mathfrak{u}(\zeta)T^* = \mathfrak{u}(\zeta^k)T^* = T^*$, equals $k \cdot \mathrm{Id}$. On the level of sections, this isomorphism sends a Hermitian connection ∇ in ζ onto $\nabla^{(k)}$ (see §4.13).

(xi) The projection $\mathcal{C}(\beta) \to \mathcal{C}_\iota(\beta)$ in (7.15) is given by $\nabla \mapsto \nabla^{[\iota^\perp]} \oplus \nabla^{[\iota]}$, the summands being defined as in (ix) of §6.6 with the aid of orthogonal projections of β onto $\iota^\perp, \iota$. Now one can easily derive (7.15) using the fact that $\nabla - (\nabla^{[\iota^\perp]} \oplus \nabla^{[\iota]})$ is a section of $[\mathfrak{su}_\iota(\beta)]^\perp T^*$.

(xii) Since the first Chern class c_1 classifies complex line bundles over $\mathcal{M}$ topologically, a cubic root $\lambda^{1/3}$ of the electromagnetism bundle λ exists if and only if $c_1(\lambda)$ is divisible by 3 in $H^2(\mathcal{M}, \mathbb{Z})$. If $\lambda^{1/3}$ exists, it is unique only in the case where $H^2(\mathcal{M}, \mathbb{Z})$ has no elements of order 3. Even under the assumption that $\lambda^{1/3}$ exists and is topologically unique, it cannot be obtained from λ by an explicit (functorial) construction.

(xiii) In weak interactions, the d and s quarks (and, possibly, the b quark) appear not in their pure form, but rather as specific *quark mixtures* d′, s′ (and, possibly, b′), so that the summands in (7.10) actually represent ν_e, e, $\overline{\mathrm{d}}'$, u. The same applies to the other fermion generations. See §§9.2, 9.3.

(xiv) The coupling constant for the SU(5) grand unified theory is $C_0 = G_{\mathrm{gu}} = r_0^{-1/2}$ with r_0 as in (7.11), cf. (6.34). By (7.13), this leads to the coupling constants $C_{\mathrm{ew}} = \sqrt{0.4}\,G_{\mathrm{gu}}$, $\tilde{C}_{\mathrm{ew}} = \sqrt{0.24}\,G_{\mathrm{gu}}$ for the corresponding electroweak model. According to (6.37) and (7.14), the electron charge then is given by $q = -\sqrt{0.15}\,\hbar G_{\mathrm{gu}}$, while the electromagnetism coupling constant is, by (6.36), $\tilde{C}_{\mathrm{em}} = \sqrt{0.15}\,G_{\mathrm{gu}}$. On the other hand, for the strong interaction bundle ρ defined in the SU(5) theory by (7.6), $\mathfrak{su}(\rho)$ carries a fibre metric induced by (7.11) via the chain of injective bundle morphisms $\mathfrak{su}(\rho) = \mathfrak{su}(\bar\rho) = \mathfrak{su}(\lambda^{-1/3}\bar\rho) = \mathfrak{su}(\iota^\perp) \to \mathfrak{su}(\beta)$, the last three of which are given by $a \mapsto \mathrm{Id}_{\lambda^{-1/3}} \otimes a$, (7.8), and $b \mapsto \mathrm{Id}_\iota \oplus b$, along with $\beta = \iota \oplus \iota^\perp$. In view of (6.34), this leads to the strong-interaction coupling constant $C_{\mathrm{strong}} = \sqrt{0.6}\,G_{\mathrm{gu}}$.

References: Mohapatra 1986; Leite Lopes 1981.

8 Unitary Symmetries of Hadrons*

This chapter deals with further consequences of the quark model, such as a simple explanation of the tendency, displayed by hadron species in nature, to form groupings known as $SU(n)$ *multiplets*, $2 \leq n \leq 6$. The latter property is known as the $SU(n)$ *symmetry* and includes, as a special case with $n = 2$, the isospin symmetry discussed in Chapter 2.

8.1 Quarks and Quantum Numbers

The values assumed for quarks by the discrete particle invariants listed in §2.14 can be described as follows.

All (anti)quarks have spin $s = \frac{1}{2}$, so that they are fermions, with statistics -1. The (conventional) parity is positive for quarks and negative for antiquarks. (See §6.2.)

Quarks and antiquarks form two isospin doublets $\{u, d\}, \{\bar{u}, \bar{d}\}$ and eight singlets $s, c, b, t, \bar{s}, \bar{c}, \bar{b}, \bar{t}$. Thus, the values of the isospin I are $I = \frac{1}{2}$ for $u, d, \bar{u}, \bar{d}$ and $I = 0$ for all other (anti)quarks (cf. (2.13)).

The C-parity ϵ_C, G-parity ϵ_G and helicity are, for obvious reasons, not defined for (anti)quarks, while the values assigned to them by the isospin class $(-1)^{2I}$ are determined by those of I, listed above.

The remaining quantum numbers discussed in §2.14 are the additive generalized charges. It is sufficient to describe them for quarks (proper), as their values for antiquarks then will be the corresponding opposite numbers. Thus, we have the electric charge $Q = \frac{2}{3}$ for u, c, t and $Q = -\frac{1}{3}$ for d, s, b (see §6.2), the lepton numbers $L_e = L_\mu = L_\tau = L = 0$ and the (non-integral!) baryon number $B = \frac{1}{3}$ for all quarks, the isospin component I_3 equal to $\frac{1}{2}$ for u, $-\frac{1}{2}$ for d, and 0 for s, c, b, t (cf. the values of the isospin I above), while the hypercharge Y, according to its definition (2.15), has the values $\frac{1}{3}$ for u, d, $\frac{4}{3}$ for c, t and $-\frac{2}{3}$ for s, b (see also Remark 8.1.1 below). Finally, the strangeness S, charm C, bottomness b and topness t can be used to distinguish quark flavors, as each of them vanishes for all quarks except one, the nonzero values being $S = -1, C = 1, b = -1$ and $t = 1$ for the s, c, b and t quark, respectively.

Using the property of S, C, b and t just stated, one can easily verify (see Remark 8.1.2) that the quark content of any given hadron species is uniquely determined by its generalized charges Q, B, I_3, Y, S, C, b, t, unless they all vanish. However, their values do not specify *how* the constituent quark flavors are put together. On the other hand, all additive generalized charges that are conserved in strong interactions (§2.13, §2.14) are automatically equal to zero for each pair $u\bar{u}, \ldots, t\bar{t}$ consisting of a quark and its own antiquark, as well as for any strictly neutral hadron (meson), cf. §1.6. The question of finding the quark content of strictly neutral mesons is therefore more difficult than for other hadrons and cannot be solved just by using their quantum numbers. Note that isospin can only be used to arrive at a partial answer, cf. Remark 8.1.3, while spin and parity yield no information at all, as they fail to distinguish quark flavors. The final answer can only be found with the aid of additional considerations involving flavor symmetries. Many strictly neutral mesons turn out to be *mixtures* of several of the pairs $u\bar{u}, \ldots, t\bar{t}$. (See (vii) of §8.4.)

The quark contents of most known hadrons, listed in §6.2 for the lightest ones, are determined as above (up to ambiguities involving strictly neutral mesons), by their generalized charges and isospin. The resulting picture is consistent with hadron quantum numbers in the sense that it not only renders correct values of all additive generalized charges, summed over the constituent quarks, but also leads to agreement with the values of isospin, spin and parity, which are not uniquely determined by the quark content alone (see (2.23), (2.24)). This is another powerful argument supporting the quark theory.

Remark 8.1.1. With the above assignments of quantum numbers to quarks, the Gell-Mann–Nishijima relation (2.22) is obviously valid for all quark flavors and hence, by additivity, for all hadrons. Thus, since the hypercharge Y is defined by (2.15), i.e., (2.16), we obtain

$$Y = B + S + C + b + t. \tag{8.1}$$

In the physical literature one often uses other definitions of Y, *not equivalent* to (8.1), and hence not satisfying (2.16). They are based on extending formula (2.20) to heavier hadrons by adding to its right-hand side combinations of C, b and t different from the one in (8.1).

Remark 8.1.2. Let A be the set consisting of all unordered quark triples, antiquark triples (with possible repetitions, i.e., "multiplicities") and quark-antiquark pairs, where by an (anti)quark we mean an (anti)quark flavor (species), and let $f : A \to \mathbb{Z}^6$ be given by $f = (Q, B, S, C, b, t)$, the value assigned by each generalized charge $Q, \ldots, t$ to any triple (pair) being the sum of its values over the constituent (anti)quarks. By definition, A is decomposed into three disjoint subsets, corresponding to baryon-proper, antibaryon, and meson quark contents (§6.2), which are distinguished by f as they are characterized by B being $1, -1$, and 0, respectively. The f-preimage of $(0, \ldots, 0)$ consists of the six pairs $u\bar{u}, \ldots, t\bar{t}$, while f restricted to $A \backslash \{u\bar{u}, \ldots, t\bar{t}\}$ is bijective onto its image, as one easily verifies using first

B, then S, C, b, t, and finally Q. Thus, the generalized charges Q, B, S, C, b, t determine the quark contents of all hadrons except strictly neutral mesons. (The hypercharge Y and isospin component I_3 can be omitted here, as they are expressed by (8.1) and (2.22).) The image of f, i.e., the collection of possible values of $Q, \ldots, t$ for hadrons, is subject to many obvious restrictions, such as $|Q| \leq 2$, $|B| \leq 1$, $|S| + |C| + |b| + |t| \leq 3$, etc.

Remark 8.1.3. The isospin symmetry amounts to dividing the set of all hadron species into mutually disjoint multiplets, each of which consists of particles sharing many properties, e.g., having the same spin, parity, quantum numbers B, Y, S, C, b, t, and almost equal masses. The above description of quark multiplets indicates that the quark contents of all hadrons forming a given multiplet should become identical if we assume that the u and d quarks coincide, ignoring the difference between their electric charges. (See §8.5 for a more formal discussion of isospin in the quark model.) As a consequence, the quark contents of the strictly neutral mesons π^0, ρ^0, δ^0 (§6.2) are expected to involve the pairs $u\bar{u}, d\bar{d}$, but no other (anti)quark flavor, in view of the quark contents of their multiplet partners $\pi^\pm, \rho^\pm, \delta^\pm$ determined as in Remark 8.1.2.

Remark 8.1.4. The Gell-Mann–Nishijima relation (2.22), instead of being a mysterious equality valid for all hadrons, now is a *consequence* of the quark model (Remark 8.1.1). More generally, any linear relation involving additive generalized charges will hold for all hadrons, provided that it does for all the quark flavors u, d, s, c, b, t. The fact that *some* relation of this type happens to be satisfied by all quarks is no surprise, as the seven quantities involved in (2.22), applied to six quark flavors, lead to a 7×6 rational matrix Φ whose rows must be linearly dependent over $\mathbb{Q}$. (As rank $\Phi = 6$, no further such relations exist.)

References: Leite Lopes 1981; Ryder 1986.

8.2 Representations of SU(n)

Throughout this section, by a *representation* we mean a continuous, linear representation of the Lie group (or algebra) in question in a finite dimensional complex vector space.

For each integer $n \geq 2$, compactness of the special unitary group SU(n) implies that every representation of SU(n) is a direct sum of irreducible subrepresentations. Let $\mathcal{R}_n$ be the set of equivalence classes of irreducible representations of SU(n). For any $\alpha_1, \ldots, \alpha_k \in \mathcal{R}_n$, we denote $\mathcal{S}(\alpha_1, \ldots, \alpha_k)$ the finite subset of $\mathcal{R}_n$ consisting of all irreducible subrepresentations (summands) of the tensor product $\alpha_1 \otimes \ldots \otimes \alpha_k$. For more details, see Remarks 8.2.1, 8.2.5, 8.2.6 below.

The center $\mathbb{Z}_n \cdot \mathrm{Id}$ of SU(n) is isomorphic to the cyclic group

$$\mathbb{Z}_n = \{z \in \mathbb{C} : z^n = 1\},$$

which we always describe using multiplicative notation. The image $F = \varphi(e^{2\pi i/n}\mathrm{Id})$ of the generator $e^{2\pi i/n}\mathrm{Id}$ of $\mathbb{Z}_n \cdot \mathrm{Id}$ under any representation φ of SU(n) commutes with φ and satisfies $F^n = \mathrm{Id}_W$, W being the representation space. If F itself happens to be a multiple of Id_W, one defines the *class* $\mathrm{cl}(\varphi)$ of the representation φ to be the unique complex number with $F = \mathrm{cl}(\varphi) \cdot \mathrm{Id}_W$. Obviously, $\mathrm{cl}(\varphi) \in \mathbb{Z}_n$. Furthermore, by Schur's lemma, $\mathrm{cl}(\varphi)$ is well-defined whenever φ is irreducible. Thus, representations of SU(n) with a well-defined class are precisely those the irreducible summands of which are all of the same class.

Restricted to irreducible representations, the class gives rise to a mapping $\mathrm{cl} : \mathcal{R}_n \to \mathbb{Z}_n$. It is clear that

$$\mathcal{S}(\alpha_1, \ldots, \alpha_k) \subset \mathrm{cl}^{-1}(\prod_{j=1}^{k} \mathrm{cl}(\alpha_j)) \tag{8.2}$$

whenever $\alpha_1, \ldots, \alpha_k \in \mathcal{R}_n$, since, under tensor multiplication of representations, the existence of class is preserved and the class mapping is multiplicative (cf. §3.2).

Replacing the complex structure of the representation space W by its conjugate (§1.6), one obtains the *conjugate representation* $\overline{\varphi}$ of any representation φ of SU(n), which gives rise to an involutive mapping $(\ )^- : \mathcal{R}_n \to \mathcal{R}_n$. Obviously,

$$\mathrm{cl}(\overline{\varphi}) = \overline{\mathrm{cl}(\varphi)} = \frac{1}{\mathrm{cl}(\varphi)} \tag{8.3}$$

whenever $\mathrm{cl}(\varphi)$ exists. Since SU(n) is compact, φ leaves invariant a Hermitian inner product $\langle , \rangle$ in W and the isomorphism $\overline{W} \to W^*$ induced by $\langle , \rangle$ leads to the identification

$$\overline{\varphi} = \varphi^*, \tag{8.4}$$

i.e., establishes an equivalence between $\overline{\varphi}$ and the representation φ^* of SU(V) in W^*, *dual* to φ.

The set $\mathcal{R}_n$ is countably infinite, and it is convenient to identify it with the Cartesian product $(\frac{1}{n}\mathbb{Z}_+)^{n-1}$ of $n-1$ copies of $\frac{1}{n}\mathbb{Z}_+ = \{y \geq 0 : ny \in \mathbb{Z}\}$. The specific identification we are going to use can be characterized as follows. Let V be an n-dimensional complex vector space with a fixed Hermitian inner product $\langle , \rangle$, and let SU(V) $\cong$ SU(n) be the group of all linear isometries of V with determinant 1. A mapping

$$(\frac{1}{n}\mathbf{Z}_+)^{n-1} \to \mathcal{R}_n \tag{8.5}$$

then is uniquely determined by the requirement that it be bijective and send each $(s_1, \ldots, s_{n-1})$ onto the equivalence class of an irreducible subrepresentation of the obvious action of $SU(V)$ (or $SU(n)$, cf. Remark 8.2.2) in the space

$$W_{s_1,\ldots,s_{n-1}} = \prod_{j=1}^{n-1} S^{ns_j}(\Lambda^j V), \tag{8.6}$$

where the product corresponds to tensor multiplication and S^k stands for the kth symmetric power. (See Remarks 8.2.1, 8.2.5.)

Identifying $\mathcal{R}_n$ and $\left(\frac{1}{n}\mathbf{Z}_+\right)^{n-1}$ by (8.5), we may regard cl and $(\ )^-$ as mappings of $\left(\frac{1}{n}\mathbf{Z}_+\right)^{n-1}$ into $\mathbf{Z}_n$ or itself. They are then given by

$$\mathrm{cl}(s_1 \ldots, s_{n-1}) = \exp(2\pi\mathrm{i} \sum_{j=1}^{n-1} js_j) \tag{8.7}$$

and

$$(s_1, \ldots, s_{n-1})^- = (s_{n-1}, \ldots, s_1). \tag{8.8}$$

In fact, the inclusions

$$\Lambda^k V, \ S^k V \subset \bigotimes^k V \tag{8.9}$$

(see Remark 8.2.7) identify (8.6) with a subspace of an appropriate tensor power of V, so that (8.7) follows from multiplicativity of cl under the tensor product along with the obvious relation $\mathrm{cl}(\varphi) = e^{2\pi\mathrm{i}/n}$ for the standard (identity) representation φ of $SU(V)$ in V. On the other hand, (8.10) below and (8.6) give rise to an $SU(V)$-equivariant isomorphism $\overline{W}_{s_1,\ldots,s_{n-1}} \to W_{s_{n-1},\ldots,s_1}$. In view of its uniqueness property, (8.5) therefore coincides with the map $(s_1, \ldots, s_{n-1}) \mapsto (s_{n-1}, \ldots, s_1)$ followed by (8.5) and then by $(\ )^-$, which implies (8.8).

Given an irreducible representation φ of $SU(n)$ (or $SU(V)$, with V as above), we refer to $(s_1, \ldots, s_{n-1}) \in \left(\frac{1}{n}\mathbf{Z}_+\right)^{n-1}$ corresponding to φ under (8.5) as the *spin* of φ. This is consistent with the terminology previously introduced for $n = 2$ (see Remark 8.2.3).

In the following remarks V stands, as before, for a complex vector space of dimension $n \geq 2$ with a fixed Hermitian inner product.

Remark 8.2.1. For each $(s_1, \ldots, s_{n-1}) \in \left(\frac{1}{n}\mathbf{Z}_+\right)^{n-1}$, (8.6) contains a unique $SU(V)$-invariant subspace $V^{(s_1,\ldots,s_{n-1})}$ the representation of $SU(V)$ in which is irreducible and corresponds to $(s_1, \ldots s_{n-1})$ under (8.5) (i.e., has the spin $(s_1, \ldots, s_{n-1})$). Explicit descriptions of $V^{(s_1,\ldots,s_{n-1})}$, as well as $\mathcal{S}(\alpha_1,\ldots,\alpha_k)$ for $\alpha_1,\ldots,\alpha_k \in \mathcal{R}_n$, are in general complicated. We will, however, only need them in a few special cases, discussed in Remarks 8.2.5 and 8.2.6 (see also Remarks 8.2.10, 8.2.11).

Remark 8.2.2. Any linear isometry $\Phi : V \to \mathbb{C}^n$ leads to an isomorphism $\mathrm{SU}(V) \cong \mathrm{SU}(n)$ and, consequently, to an identification of the set of equivalence classes of irreducible representations of $\mathrm{SU}(V)$ with $\mathcal{R}_n$, which obviously does *not* depend on Φ.

Remark 8.2.3. For $n = 2$, the above definition of spin coincides with that in §2.1. This is immediate from our characterization of (8.5) along with the fact that *the representation of* $\mathrm{SU}(V) \cong \mathrm{SU}(2)$ *in the symmetric power* $S^k V$ *is irreducible and its spin (in the sense of §2.1) equals* $\frac{k}{2}$. The last assertion follows, for $k = 1$, from the dimension-spin relation in §2.1, so that, given a fixed unit $X \in \mathfrak{su}(V)$, we can select $v \in V \smallsetminus \{0\}$ with $Xv = \frac{1}{2}iv$ (cf. (2.2), (2.3)). The case of any $k \geq 1$ now is easily obtained from the irreducibility criterion (§2.1) applied to X and the kth symmetric power of v, along with the obvious relation $\dim S^k V = k + 1$ (whenever $\dim V = 2$).

Remark 8.2.4. Let φ_j be the obvious representation of $\mathrm{SU}(V)$ in $\Lambda^j V$, $j = 1,\ldots,n$. Since φ_n is trivial, the $\mathrm{SU}(V)$-equivariant pairing $\Lambda^j V \otimes \Lambda^{n-j} V \to \Lambda^n V \cong \mathbb{C}$ of exterior multiplication establishes an equivalence between φ_{n-j} and φ_j^* and, by (8.4), also between φ_{n-j} and $\overline{\varphi}_j$. In other words, for $j = 0,\ldots,n$, we have an $\mathrm{SU}(V)$-equivariant isomorphism

$$\Lambda^{n-j}V = \overline{\Lambda^j V}. \tag{8.10}$$

Remark 8.2.5. Some particular examples of the irreducible representation spaces $V^{(s_1,\ldots,s_{n-1})}$ with spins $(s_1,\ldots s_{n-1})$ for the group $\mathrm{SU}(V) \cong \mathrm{SU}(n)$ (Remark 8.2.1), are provided by

$$S^{ns}V = V^{(s,0,\ldots,0)}, \qquad s \in \frac{1}{n}\mathbb{Z}_+ \tag{8.11}$$

and, for $k = 0, 1,\ldots,n$,

$$\Lambda^k V = V^{(s_1,\ldots,s_{n-1})} \quad \text{with} \quad s_j = \frac{1}{n}\delta_{jk}, \tag{8.12}$$

including

$$V = V^{(\frac{1}{n},0,\ldots,0)}, \qquad \mathbb{C} = V^{(0,\ldots,0)}. \tag{8.13}$$

Furthermore, let $V^{k,\ell}$ be the kernel of the mapping $\Lambda^k V \otimes \Lambda^\ell V \to \Lambda^{k+1}V \otimes \Lambda^{\ell-1}V$ characterized by $(v_1 \wedge \ldots \wedge v_k) \otimes (w_1 \wedge \ldots \wedge w_\ell) \mapsto \sum_{j=1}^{\ell}(-1)^{j-1}(v_1 \wedge \ldots \wedge v_k \wedge w_j) \otimes (w_1 \wedge \ldots \wedge w_{j-1} \wedge w_{j+1} \wedge \ldots \wedge w_\ell)$, where k, ℓ are integers with $n \geq k > \ell \geq 1$. Then

$$V^{k,\ell} = V^{(s_1,\ldots,s_{n-1})} \quad \text{with} \quad s_j = \delta_{kj} + \delta_{\ell j}. \tag{8.14}$$

Note that, in view of (8.9) we have, for $k = 2, \ldots, n$,

$$V^{k,1} = \text{Ker} \wedge, \tag{8.15}$$

$\wedge : \Lambda^k V \otimes \Lambda^1 V \to \Lambda^{k+1} V$ being the (surjective) exterior multiplication, which establishes the $SU(V)$-equivariant isomorphism

$$\Lambda^k V \otimes V = V^{k,1} \oplus \Lambda^{k+1} V. \tag{8.16}$$

Consequently, if $2 \le k \le n = \dim V$,

$$\dim V^{k,1} = k \binom{n+1}{k+1}. \tag{8.17}$$

and

$$V^{n,1} \cong V. \tag{8.18}$$

Remark 8.2.6. With $V^{k,1}$ given by (8.15), we have the $SU(V)$-equivariant, natural isomorphisms

$$V \otimes V = S^2 V \oplus \Lambda^2 V, \tag{8.19}$$

$$V \otimes V \otimes V = S^3 V \oplus V^{2,1} \oplus V^{2,1} \oplus \Lambda^3 V, \tag{8.20}$$

$$V \otimes \overline{V} = V^{n-1,1} \oplus \mathbb{C} \quad \text{if} \quad n = \dim V \ge 3. \tag{8.21}$$

In fact, (8.19) is obvious form (8.9). Moreover, the surjective *symmetrization mapping* $\Sigma : S^2 V \otimes V \to S^3 V$ with $\Sigma((u \odot v) \otimes w) = u \odot v \odot w$ (where $\odot$ is the symmetric product) leads to the isomorphism $S^2 V \otimes V = S^3 V \oplus \text{Ker}\,\Sigma$. On the other hand, the linear map $T : S^2 V \otimes V \to V^{2,1}$ with $T((u \odot v) \otimes w) = (w \wedge u) \otimes v + (w \wedge v) \otimes u$, restricted to $\text{Ker}\,\Sigma$ is an isomorphism (with the inverse $V^{2,1} \to \text{Ker}\,\Sigma$ characterized by $(w \wedge u) \otimes v \mapsto \frac{1}{3}[(u \odot v) \otimes w - (w \odot v) \otimes u]$). Thus, $S^2 V \otimes V = S^3 V \oplus V^{2,1}$ and so (8.20) follows from $V \otimes V \otimes V = (S^2 V \oplus \Lambda^2 V) \otimes V$ (cf. (8.19)) along with (8.16) for $k = 2$. Finally, (8.21) is immediate from (8.16) and (8.10) with $k = n-1$ and $j = 1$.

Note that, under the identification $\overline{V} = V^*$, (8.21) coincides with the decomposition

$$V \otimes V^* = \text{End}\,V = \mathfrak{sl}(V) \oplus \mathbb{C} = \text{Ker}\,(\text{Trace}) \oplus \mathbb{C} \cdot \text{Id}. \tag{8.22}$$

Remark 8.2.7. Our notational conventions about symmetric and exterior multiplications are such that $k!\, v_1 \odot \ldots \odot v_k = \sum_\varphi v_{\varphi_1} \otimes \ldots \otimes v_{\varphi_k}$ and $k!\, v_1 \wedge \ldots \wedge v_k = \sum_\varphi (\text{sgn}\,\varphi) v_{\varphi_1} \otimes \ldots \otimes v_{\varphi_k}$ for $v_j \in V$, both summations being over all permutations φ of $\{1, \ldots, k\}$. Thus, we have (8.9) and projections of $\bigotimes^k V$ onto $S^k V$, $\Lambda^k V$ can be defined by assigning $v_1 \odot \ldots \odot v_k$ and

$v_1 \wedge \ldots \wedge v_k$, respectively, to $v_1 \otimes \ldots \otimes v_k$. Moreover, Σ in Remark 8.2.6 and $\wedge$ in (8.15) then also become projections, so that the projections onto their kernels equal $\mathrm{Id} - \Sigma$ and $\mathrm{Id} - \wedge$, respectively.

Remark 8.2.8. In view of Remark 8.2.7 and the discussion in Remark 8.2.6, the projections onto the individual summands in (8.19) – (8.21) can be characterized as follows.

(i) In (8.19), they assign $v \odot w \in S^2 V$ and $v \wedge w \in \Lambda^2 V$, respectively, to $v \otimes w \in V \otimes V$.

(ii) In (8.20), they send $v \otimes w \otimes u \in V \otimes V \otimes V$ onto $v \odot w \odot u \in S^3 V$, $v \otimes (w \odot u) - v \odot w \odot u \in \mathrm{Ker}\, \Sigma \cong V^{2,1}$, $v \otimes (w \wedge u) - v \wedge w \wedge u \in V^{2,1}$ and $v \wedge w \wedge u \in \Lambda^3 V$, respectively.

(iii) With (8.21) replaced by (8.22), the projections of $\mathrm{End}\, V$ onto $\mathfrak{sl}(V)$ and $\mathbb{C}$ are given by $F \mapsto F - \frac{1}{n}(\mathrm{Trace}\, F)\mathrm{Id}$ and $F \mapsto \frac{1}{n}\mathrm{Trace}\, F$, respectively, where $n = \dim V$. The isomorphism $V \otimes V \to \mathrm{End}\, V$ sends $v \otimes w$ onto A with $Au = \langle u, w \rangle v$ for $u \in V$ (which is antilinear in $w \in V$ and hence becomes linear when w is regarded as an element of $\overline{V}$).

Remark 8.2.9. Due to the repeated summand $V^{2,1}$, the decomposition (8.20) is not unique, despite its uniqueness up to equivariant isomorphisms. According to Remarks 9.1.1, 9.1.3, nontrivial irreducible SU(V) invariant subspaces in the span of $\mathrm{Ker}\, \Sigma \cong V^{2,1}$ and $V^{2,1}$ are classified by $\mathbb{C}P^2$. Let pr be the projection of $V \otimes V \otimes V$ onto $V^{2,1}$ and let $\Phi : V^{2,1} \to \mathrm{Ker}\, \Sigma$ be a fixed equivariant isomorphism. The projections of $V \otimes V \otimes V$ onto the subspaces just mentioned, other than $\mathrm{Ker}\, \Sigma$, are easily seen to be of the form $\mathrm{pr} + z\Phi \circ \mathrm{pr}$ with $z \in \mathbb{C}$.

Remark 8.2.10. For $\alpha_1, \ldots, \alpha_k \in \mathcal{R}_n$, it is often much easier to describe some finite sets containing $\mathcal{S}(\alpha_1, \ldots, \alpha_k)$ rather than $\mathcal{S}(\alpha_1, \ldots, \alpha_k)$ itself. For instance, let $\lambda_1, \ldots, \lambda_n$ be real numbers with

$$\sum_{j=1}^{n} \lambda_j = 0. \tag{8.23}$$

A matrix a in the Lie algebra $\mathfrak{su}(n)$ with the eigenvalues $i\lambda_1, \ldots, i\lambda_n$ then is uniquely determined up to a conjugation under SU(n), and hence so is its image $\tau(a)$ under the representation τ of $\mathfrak{su}(n)$ induced by any given representation φ of SU(n). Thus, the largest eigenvalue $F(\varphi) = F_{\lambda_1, \ldots, \lambda_n}(\varphi)$ of $-i\tau(a)$ is an invariant of φ, depending on $\lambda_1, \ldots, \lambda_n$ with (8.23), but not on the choice of a. (Since SU(n) is compact, the spectrum of $\tau(a)$ lies in $i\mathbb{R}$.) Clearly, $F(\varphi_1 \oplus \ldots \oplus \varphi_k) = \max_{1 \leq j \leq k} F(\varphi_j)$ and

$$F(\varphi_1 \otimes \ldots \otimes \varphi_k) = \sum_{j=1}^{k} F(\varphi_j)$$

for any representations $\varphi_1, \ldots, \varphi_k$ of SU(n). Restricted to irreducible representations, $F = F_{\lambda_1, \ldots, \lambda_n}$ therefore becomes a function $F : \mathcal{R}_n \to \mathbb{R}$ with

$$\mathcal{S}(\alpha_1, \ldots, \alpha_k) \subset F^{-1}((-\infty, \sum_{j=1}^{k} F(\alpha_j)]) \tag{8.24}$$

whenever $\alpha_1, \ldots, \alpha_k \in \mathcal{R}_n$.

Thus, choosing $(\lambda_1, \ldots, \lambda_n)$ to be $(1 - n, 1, \ldots, 1)$ (or $(n-1, -1, \ldots, -1)$) and identifying each $\alpha \in \mathcal{R}_n$, via (8.5), with its spin $(s_1, \ldots, s_{n-1}) \in \left(\frac{1}{n}\mathbb{Z}_+\right)^{n-1}$, we obtain $F(s_1, \ldots, s_{n-1}) = n \sum_{j=1}^{n-1} j s_j$ (or $F(s_1, \ldots, s_{n-1}) = n \sum_{j=1}^{n-1} (n - j) s_j$). Consequently, $\mathcal{S}(\alpha_1, \ldots, \alpha_k)$ satisfies three inclusion relations, namely (8.24) with these two choices of F, and (8.2) (with (8.7)).

Remark 8.2.11. For any $\alpha_1, \ldots, \alpha_k \in \mathcal{R}_n$, $\mathcal{S}(\alpha_1, \ldots, \alpha_k)$ contains the element $\alpha_1 + \ldots + \alpha_k$ with *multiplicity one*, i.e., as a non-repeated summand of $\alpha_1 \otimes \ldots \otimes \alpha_k$. Here $+$ stands for ordinary addition in the lattice semigroup $\left(\frac{1}{n}\mathbb{Z}_+\right)^{n-1}$, identified with $\mathcal{R}_n$ via (8.5).

References: Zhelobenko 1973; Huang 1982; Lichtenberg 1978.

8.3 Flavor Symmetries

The isospin symmetry, discussed in §2.9 (and briefly summarized in Remark 8.1.3), can be made consistent with the quark model in a manner that is unique, yet seemingly strange. Specifically, the way quarks are divided into isospin multiplets determines, and is determined by, the multiplet structure of hadrons, through an intuitively clear mechanism, which has a rigorous description in terms of group representations (see below). As a consequence, the isospin symmetry for quarks *must* be as described in §8.1, so that the u and d quarks form a doublet, while the remaining quark flavors are isospin singlets (similarly for antiquarks). Thus, it is natural to ask *what* makes u and d so different from the other quarks.

A possible explanation is that, by (6.2), they are the two *lightest* quark flavors. This in turn suggests that, unless some particular significance is attached to the number two, there may as well be a more general, but somehow similar *n-flavor symmetry* involving the n lightest quark species, for each integer n with $2 \leq n \leq N$, where N is the total number of quark flavors in nature. (One currently knows that $N \geq 5$, and believes that $N = 6$, cf. Remark 6.2.4.) Experimental evidence shows that the n-flavor symmetry actually exists in nature, its reality being manifested by a conservation law; see Remark 8.3.9 below.

Before proceeding further, it is useful to introduce the notion of a *generalized hadron*, by which we mean any composite-particle species built from quarks. The manner in which the constituent quarks are put together is purely formal, as in (1.1), (1.2), so that the resulting composite objects are neither assumed, nor in general expected to be physically stable. Thus, beside quarks and genuine hadrons (§6.2), the class of generalized hadrons also contains bound states of hadrons (including nuclei), and what one may think of as configurations formed by several interacting hadrons at the instant of their mutual collision. (See Remark 8.3.8.)

We can now derive some basic properties of the n-flavor symmetry (also called SU(n) *symmetry*, for reasons given in the next two paragraphs) by extrapolating what we know about isospin ($n = 2$) to higher values of n. In this way, the n-flavor symmetry amounts to dividing the set of all generalized hadrons into pairwise disjoint SU(n) *multiplets*. This includes, for $n = 2$, the (isospin) multiplets formed by hadrons (§2.9) and quarks (§8.1). Antiparticle formation leaves this partition invariant, sending each SU(n) multiplet onto the corresponding *antimultiplet*. As in the case of isospin, the set of all quark flavors and the class of hadrons are both unions of SU(n) multiplets and, in particular, the lightest n quarks belong to an n-tuplet, while the remaining quark flavors form SU(n) singlets. (Similarly for antiquarks.) Furthermore, the n-flavor symmetry is a generalized spin symmetry as in (b) of §1.9, since so is isospin (§2.10). Thus, if a given SU(n) multiplet consists of k generalized hadrons, their common generalization is represented by a vector bundle of the form

$$\eta = \eta_0 \otimes (\mathcal{M} \times V) \tag{8.25}$$

over the spacetime $(\mathcal{M}, g)$, where V is a complex vector space of dimension k, called the SU(n) *space* of the multiplet and carrying some additional geometric structure. By splitting V into lines in a manner compatible with its geometry, we obtain a direct-sum decomposition of η into k subbundles corresponding to the member particles and isomorphic to η_0. Consequently, all particles in a fixed SU(n) multiplet are described by a common natural bundle and so they all have the same spin and parity (§§1.14, 1.16, 1.17), while other properties they may be expected to share due to analogy with isospin (Remark 8.1.3) include some quantum numbers and a common mass range (Remarks 8.3.4, 8.3.5).

An obvious choice of the SU(n) space for SU(n) *singlets* is $V = \mathbb{C}$. Letting V_n be the SU(n) space for the SU(n) multiplet of n lightest quarks, we have $\dim V_n = n$ and, in particular, $V_2 = S$ with $(S, \Omega, \langle,\rangle)$ as in §2.10. Thus, it is natural to require that the geometry of each V_n consist of a Hermitian inner product $\langle,\rangle$ and a compatible $\Omega \in \Lambda^n V_n^*$, i.e., of an SU($n$)-structure (cf. (iii) of §5.2). On the other hand, under antiparticle formation (§1.6), η in (8.25) is replaced by $\overline{\eta} = \overline{\eta}_0 \otimes (\mathcal{M} \times \overline{V})$, so that the antimultiplet of ·an SU(n) multiplet with the SU(n) space V is always represented by the conjugate space $\overline{V}$. As a consequence, $\overline{V}_n$ and $\overline{\mathbb{C}} = \mathbb{C}$ are the SU(n) spaces of the "light" and "heavy" antiquarks, respectively, which completes the description of SU(n) spaces for all (anti)quarks.

The composite-particle formulae (1.1), (1.2) applied to factors of the form (8.25), show that the SU(n) space for an object obtained by putting together generalized hadrons with SU(n) spaces $W_1, \ldots, W_k$, has to be a "naturally defined" subspace (summand, i.e., a projection image) of $W_1 \otimes \ldots \otimes W_k$. The way hadrons are composed of quarks (§6.2) now implies that the SU(n) space V of each hadron is of this type with $k \leq 3$, each factor W_j being one of $V_n, \overline{V}_n$ or $\mathbb{C}$, while "naturality" of V is related to the geometric structure $\Omega, \langle,\rangle$ of V_n. In other words, V has to be invariant under the group SU(V_n) $\cong$ SU(n) (cf. §8.2) acting on V_n, $\overline{V}_n$, $\mathbb{C}$ and their tensor products in the obvious way. As in the case of isospin, we expect the resulting representation of SU(V_n) $\cong$ SU(n) in V to be irreducible. Consequently, we may define the SU(n) *spin* $(s_1, \ldots, s_{n-1}) \in \left(\frac{1}{n}\mathbb{Z}_+\right)^{n-1}$ and SU(n) *class* $\exp(2\pi i \sum_{j=1}^{n-1} j s_j) \in \mathbb{Z}_n$ of any SU(n) multiplet of generalized hadrons (or, of each multiplet member) to be the spin (§8.2) and, respectively, class (see (8.7)) of the equivalence class of SU(n) representations associated as above (cf. Remark 8.2.2) with the SU(n) space V of the multiplet. For $n = 2$, these invariants are nothing else than the isospin $I = s_1$ and the isospin class $e^{2\pi i I} = (-1)^{2I}$ (see Remark 8.2.3 and §2.14).

Remark 8.3.1. The manner in which a given hadron is obtained as a bound state of quarks and/or antiquarks (§6.2) gives rise to several important invariants. Among them, the hadron's *quark content* indicates just which (anti)quark flavors are involved and with what multiplicities, so that it is an element of the set A defined in Remark 8.1.2. We speak of the *quark content modulo* SU(n) if only the SU(n) *multiplets* to which the constituent quarks belong are specified (again, with their multiplicities). The *quark composition* of the hadron in question is the nonzero surjective morphism (1.2), i.e.,

$$(S_0^\ell T^*)\sigma^3 \to \eta, \quad (S_0^\ell T^*)\overline{\sigma}^3 \to \eta, \quad (S_0^\ell T^*)\sigma\overline{\sigma} \to \eta \tag{8.26}$$

(notation of (5.1)), for baryons, antibaryons and mesons, respectively, where σ is the Dirac spinor bundle over the spacetime $(\mathcal{M}, g)$ representing all quark flavors, $T^* = T^*\mathcal{M}$, and η is the natural bundle over $(\mathcal{M}, g)$ in which the hadron lives (§1.16, §1.17). In particular, the quark composition determines the integer $\ell \geq 0$, called the *orbital angular momentum* contained in the given hadron, as well as its spin s and parity ϵ. (Conversely, ℓ, s and ϵ determine the quark composition (8.26) of the hadron up to a suitable "physical" equivalence. See Remark 8.7.13.) Note that (8.26) reveals nothing about the quark (flavor) content, except for distinguishing quarks form antiquarks. However, flavors can be included in (8.26), if necessary, by using flavor-labeled σ factors, such as σ_u, σ_d, etc.

Remark 8.3.2. The mechanism that holds quarks and/or antiquarks together, so as to form hadrons, is the strong interaction. Although not explicitly accounted for by the quark compositions (8.26), it can be included in (8.26) in a unique manner by replacing each σ factor with $\rho\sigma$ (where ρ is the strong-interaction bundle, cf. §6.2) and then treating the ρ, $\overline{\rho}$ factors with the morphisms $\rho^3 \to 1$, $\overline{\rho}^3 \to 1$, $\rho\overline{\rho} \to 1$ ((ii) of §5.6). For instance, instead of the first morphism in (8.26), one then obtains (setting $\zeta_\ell = S_0^\ell T^*$) the composite $\zeta_\ell(\rho\sigma)^3 = \rho^3\zeta_\ell\sigma^3 \to 1 \cdot \zeta_\ell\sigma^3 = \zeta_\ell\sigma^3 \to \eta$.

Remark 8.3.3. To account for the n-flavor symmetry, each σ factor in (8.26) has to be replaced, as in (8.25), by $\sigma \otimes (\mathcal{M} \times V)$, where V is the SU(n) space of the quark flavor in question (i.e., $V = V_n$ or $V = \mathbb{C}$). If W is the SU(n) space of the hadron represented by η, it has to be, according to the preceding discussion, an irreducible SU(V_n)-invariant direct summand of the tensor product of SU(n) spaces corresponding to the (anti)quarks involved, while the respective bundle morphism in (8.26) has to be tensored with the projection of that tensor product onto W, thus becoming valued in $\eta \otimes (\mathcal{M} \times W)$.

In this way, the hadrons with any prescribed quark content modulo SU(n) and quark composition (see Remark 8.3.1) fall into SU(n) multiplets corresponding to summands W of the tensor product just mentioned, the number of particles in each respective multiplet being $\dim W$. For more details, see (ii), (iii), (iv) of §8.4.

Remark 8.3.4. If $2 \leq n \leq 6$, each of the last $6 - n$ among the quantum numbers I_3, S, C, b, t has a common value for all members of any SU(n) multiplet of (anti)quarks, and so does the baryon number B (see §8.1). Consequently, this is also the case for SU(n) multiplets of hadrons (and generalized hadrons), as $I_3, \ldots, t, B$ are all additive and conserved in strong interactions (§2.14). Cf. Remark 8.3.7.

Remark 8.3.5. Both experimental data, and theoretical hadron-mass formulae (cf. Remark 6.2.3) indicate that the masses of hadrons forming an SU(n) multiplet, although not as nearly equal as for isospin ($n = 2$), still remain "close" in a weaker sense, characterized as follows. Let the *mass range* of an SU(n) multiplet of hadrons be the smallest interval of $\mathbb{R}$ containing the mass values of its member particles. Then the mass ranges of any two distinct SU(n) multiplets with equal spins and parities are *disjoint*.

This disjointness requirement may also be used to *define* a unique pattern of SU(n) multiplets, consistent with experimental data (see (ix) of §8.4). On the other hand, it may be derived from other theoretical arguments, such as a hadron-mass formula for $n = 3$, due to Gell-Mann and Okubo (see [Ryder 1986]).

Remark 8.3.6. Just like isospin (§2.10), the SU(n) symmetry in nature is obviously broken (approximate). Specifically, if $2 \leq n \leq 6$, the lightest n quarks can be distinguished from one another using the first $n - 1$ of the quantum numbers I_3, S, C, b, t (§8.1). According to (b) of §1.9, these quark flavors, ordered by increasing masses, correspond to complex lines $L_1, \ldots, L_n$ in V_n, forming an orthogonal direct-sum decomposiiton

$$V_n = \sum_{j=1}^{n} L_j. \tag{8.27}$$

The selection by nature of the lines L_j amounts to breaking the symmetry of V_n, i.e., to replacing the original structure group $\mathrm{SU}(n)$ by the maximal torus T^{n-1} consisting of all diagonal $n \times n$ unitary matrices of determinant 1. The corresponding T^{n-1}-structure in V_n then consists of all bases $v_1, \ldots, v_n$ such that $\Omega(v_1, \ldots, v_n) = 1$ and $|v_j| = 1$, $v_j \in L_j$ for $j = 1, \ldots, n$. The identification $\mathbb{C}^n \cong V_n$ established by such a basis leads to an isomorphism $\mathrm{SU}(n) \cong \mathrm{SU}(V_n)$ sending T^{n-1} onto the maximal torus $\mathrm{T}(V_n) \subset \mathrm{SU}(V_n)$ consisting of all unimodular linear isometries of V_n leaving each of the lines L_j invariant. (Obviously, $\mathrm{T}(V_n)$ also depends on the choice of the L_j.)

Remark 8.3.7. With L_j as in Remark 8.3.6, $n \leq 6$, one may identify the first $n - 1$ of the quantum numbers I_3, S, C, b, t (as well as B, Q, Y with (2.22), (8.1)) with skew-adjoint endomorphisms of V_n leaving each line L_j invariant and equal on L_j to i·Id times the value of the respective quantum number for the jth lightest quark. This is justified by analogous treatment of I_3 for $n = 2$ (§2.10) and by utmost convenience of the resulting identification of each of these quantum numbers with a generator of a Lie algebra action of $\mathbb{R}$ on V_n (corresponding to a linear group action of S^1). In fact, I_3, S, C, b, t and B, Q, Y now operate naturally on $V_n, \overline{V}_n$ and their tensor products (as generators of conjugates and tensor products of Lie algebra representations, cf. §3.4, §3.2) and, defining the *values* of $I_3, \ldots, Y$ to be their eigenvalues divided by i, one sees that these values, as expected (§2.13, §2.14), are additive under formation of composite objects (cf. Remark 8.3.3 above and (ii), (iii) of §8.4), and that they change sign under antiparticle formation.

Moreover, the traceless parts of the first $n-1$ among the endomorphisms I_3, S, C, b, t of V_n, obtained from them by subtracting $\frac{1}{n}\mathrm{Id}_{V_n} = \frac{3}{n}B$ times the trace of each respective endomorphism, form a basis of the Lie algebra of the maximal torus $\mathrm{T}(V_n) \cong \mathrm{T}^{n-1}$ in $\mathrm{SU}(V_n)$ (cf. Remark 8.3.6).

Remark 8.3.8. By introducing formal composite objects, such as generalized hadrons, one acquires a convenient way of phrasing conservation laws. In fact, a particle invariant which is additive (multiplicative) is conserved in a specific class of interactions if and only if its definition can be extended so as to make sense also for formal composite objects, and it then remains additive (multiplicative) when such objects are "momentarily" created from, or decay into, several ordinary particles under the influence of the given type of interactions.

More generally, formal composite objects can also be used to express what being conserved means for an invariant whose value for such an object is expected to be a *multi-valued* function of the values it assigns to the consituent particles (or decay products). An example of such an invariant is the SU(n) spin defined above for generalized hadrons where, according to the above discussion of SU(n) spaces for composite objects, the multi-valued function in question is

$$(\alpha_1, \ldots, \alpha_k) \mapsto \mathcal{S}(\alpha_1, \ldots, \alpha_k), \tag{8.28}$$

the constituent spins being identified with $\alpha_j \in \mathcal{R}_n$ via (8.5) (see §8.2). When $n = 2$, this is nothing else than the vector addition rule (2.23) for isospin.

Remark 8.3.9. Experimental evidence shows that, at least for $n = 2, 3, 4$, the SU(n) spin is conserved in strong interactions, although not as an additive invariant, but rather in the sense of the "generalized vector addition rule" (8.28).

Thus, by (8.2), the SU(n) class is conserved in strong interactions as a *multiplicative* quantity (valued in the group $\mathbb{Z}_n \subset S^1$, which we regard as multiplicative).

References: Lichtenberg 1978; Ryder 1986; Sudbery 1986.

8.4 More on SU(n) Symmetry

(i) According to §8.3 and (8.13) with $V = V_n$, the SU(n) space, SU(n) spin and SU(n) class of an (anti)quark are, respectively,
   a) V_n, $\left(\frac{1}{n}, 0, \ldots, 0\right)$ and $e^{2\pi i/n}$ for the n lightest quarks,
   b) $\overline{V}_n$, $\left(0, \ldots, 0, \frac{1}{n}\right)$ and $e^{-2\pi i/n}$ for their antiquarks,
   c) $\mathbb{C}$, $(0, \ldots, 0)$ and 1 for each remaining (anti)quark.

(ii) Let us consider all baryons with a given quark content modulo SU(n) and a specific quark composition (defined in Remark 8.3.1), and let $r \leq 3$ be the number of times one of the lightest n quarks appears among the three constituents (§6.2). By (i) and Remark 8.3.3, there are four possible cases, with SU(n) spins and SU(n) classes of the resulting SU(n) multiplets, along with the number of hadrons (baryons) in each multiplet, obtained from (8.11) – (8.14), (8.17), (8.18) with $V = V_n$ and (8.7):
   a) $r = 0$. As $\mathbb{C} \otimes \mathbb{C} \otimes \mathbb{C} = \mathbb{C}$, there is just one such baryon (an SU(n) singlet) with the SU(n) space $\mathbb{C}$, SU(n) spin $(0, \ldots, 0)$, and SU(n) class 1.
   b) $r = 1$. We have $\mathbb{C} \otimes \mathbb{C} \otimes V_n = V_n$, so that there are n such baryons, forming an SU(n) multiplet with the SU(n) space V_n, SU(n) spin $\left(\frac{1}{n}, 0, \ldots, 0\right)$ and SU(n) class $e^{2\pi i/n}$.

c) $r = 2$. By (8.19) with $V = V_n$, the n^2 baryons in question are divided into two multiplets of $\frac{n(n+1)}{2}$ and $\frac{n(n-1)}{2}$ particles, with the $SU(n)$ spaces S^2V_n, Λ^2V_n, $SU(n)$ spins $(\frac{2}{n}, 0, \ldots, 0)$, $(0, \frac{1}{n}, 0, \ldots, 0)$ for $n \geq 3$, or $1, 0$ if $n = 2$, and $SU(n)$ classes both equal to $e^{4\pi i/n}$.

d) $r = 3$. Formula (8.20) with $V = V_n$ seems to indicate that there are n^3 such baryons forming four $SU(n)$ multiplets (when $n \geq 3$), or three $SU(2)$ multiplets (for $n = 2$). However, two of these multiplets have $SU(n)$ spaces that are isomorphic (in an $SU(V_n)$-equivariant manner). Moreover, the quark contents and values of additive quantum numbers form identical patterns in both multiplets (see (v)), which hints at the possibility that the repeated summand $V_n^{2,1}$ of (8.20) might represent *just one* $SU(n)$ multiplet in nature. Experimental evidence suggests that this is actually the case. Thus, by (8.17), there are only $n^3 - 2\binom{n+1}{3} = \frac{1}{3}n(2n^2 + 1)$ baryons with stated properties, and they form three or two $SU(n)$ multiplets (depending on whether $n \geq 3$ or $n = 2$), with the $SU(n)$ spaces $S^3V_n, V_n^{2,1}$, Λ^3V_n (the last one being trivial, i.e., "absent", if $n = 2$), $SU(n)$ spins $(\frac{3}{n}, 0, \ldots, 0)$, $(\frac{1}{n}, \frac{1}{n}, 0, \ldots, 0)$, $(0, 0, \frac{1}{n}, 0, \ldots, 0)$ when $n \geq 4$, or $(1, 0), (\frac{1}{3}, \frac{1}{3}), (0, 0)$ for $n = 3$, or $\frac{3}{2}, \frac{1}{2}$ if $n = 2$, and $SU(n)$ classes all equal to $e^{6\pi i/n}$. The numbers of particles in these multiplets are, respectively, $\binom{n+2}{3}, 2\binom{n+1}{3}, \binom{n}{3}$ (the last one only for $n \geq 3$).

A similar classification of *antibaryon* $SU(n)$ multiplets is obtained by replacing each $SU(n)$ space with its conjugate and transforming the $SU(n)$ spins and classes according to (8.8) and (8.3).

(iii) As in (ii), we can also classify all mesons with a prescribed quark content modulo $SU(n)$ and a given quark composition, letting $r \leq 2$ be the number of times one of the lightest n quarks or antiquarks appears among the two constituents. This leads to four cases:

a) $r = 0$. As $\mathbb{C} \otimes \overline{\mathbb{C}} = \mathbb{C}$, there is just one such meson (an $SU(n)$ singlet), with the $SU(n)$ space $\mathbb{C}$, $SU(n)$ spin $(0, \ldots, 0)$, and $SU(n)$ class 1.

b) $r = 1$, one "light" quark involved. Since $V_n \otimes \mathbb{C} = V_n$, these n mesons form an $SU(n)$ multiplet with the $SU(n)$ space V_n, $SU(n)$ spin $(\frac{1}{n}, 0, \ldots, 0)$, and $SU(n)$ class $e^{2\pi i/n}$.

c) $r = 1$, with one "light" antiquark. As $\mathbb{C} \otimes \overline{V}_n = \overline{V}_n$, there are n such mesons, forming an $SU(n)$ multiplet with the $SU(n)$ space $\overline{V}_n$, $SU(n)$ spin $(0, \ldots, 0, \frac{1}{n})$, and $SU(n)$ class $e^{-2\pi i/n}$ (cf. (8.8), (8.3)).

d) $r = 2$. By (8.21) with $V = V_n$, the n^2 mesons in question give rise to an $SU(n)$ multiplet of $n^2 - 1$ particles and an $SU(n)$ singlet, with the $SU(n)$ spaces $V_n^{n-1,1} = \mathfrak{sl}(V_n)$ for $n \geq 3$ (cf. (8.22)) or S^2V_n if $n = 2$, and $\mathbb{C}$, respectively, $SU(n)$ spins $(\frac{1}{n}, 0, \ldots, 0, \frac{1}{n})$ and $(0, \ldots, 0)$ when $n \geq 3$, or $1, 0$ if $n = 2$, and $SU(n)$ classes both equal to 1. (Note that, for $n = 2$, (8.21) has to be replaced by (8.19), as (8.10) then implies $\overline{V} = V$.) Either of these $SU(n)$ multiplets is its own antimultiplet, and actually contains some strictly neutral mesons (see (vi)).

(iv) The projections, required in Remark 8.3.3, onto the SU(n) spaces of hadron multiplets listed in (ii), (iii) are described in Remark 8.2.8. As for the baryon multiplet in (ii)d), corresponding to either middle summand of (8.20) with $V = V_n$ (or, more generally, to any other subspace of $V_n \otimes V_n \otimes V_n$ carrying an equivalent representation of SU(V_n)), the projection used depends on which subspace one *chooses* to represent the multiplet (see Remark 8.2.9). For simplicity, the choice is usually one of the $V^{2,1}$ summands of (8.20) described in Remark 8.2.6.

Thus, a *physical* SU(n) multiplet determines the projection in (ii)d) only up to an SU(V_n)-equivariant linear isometry of the tensor-product space.

(v) Using (8.27) and the notation of (5.1), we obtain the direct-sum decompositions

$$V_n V_n = \sum_{j,k} L_j L_k, \quad V_n V_n V_n = \sum_{i,j,k} L_i L_j L_k, \quad V_n \overline{V}_n = \sum_{j,k} L_j \overline{L}_k \quad (8.29)$$

into complex lines, mutually orthogonal relative to a natural inner product (induced by $\langle , \rangle$ in V_n) in each respective tensor-product space. However, (8.29) cannot be used to describe individual hadrons (in the sense of §8.3 and (b) of §1.9), since it is not compatible with the multiplet decompositions in (ii), (iii). In fact, except for $L_j L_j$, $L_j L_j L_j$ and $L_j L_k$ with $j \neq k$, the summand lines of (8.29) are not contained in summands of (8.19) – (8.21) with $V = V_n$. On the other hand, decompositions of the latter summands into mutually orthogonal lines are obviously needed in order to exhibit the hadrons forming each given multiplet (§8.3). The only "completely natural" way of obtaining such decompositions appears to be by *projecting the lines in* (8.29) *onto the summands of* (8.19) – (8.21) *(with $V = V_n$)*.

This idea works perfectly for almost all summand lines in (8.29) (and, clearly, (8.27)), the only exceptions being those of the form $L_j \overline{L}_j$ in $V_n \overline{V}_n$. The reason is that the projections in question send elements of the other lines, all of which are tensor products, onto combinations of their images under permutations of factors (see (iv) above and Remarks 8.2.7, 8.2.8). Projection images of the lines (some of which are trivial) then are easily seen to be pairwise orthogonal or identical. Moreover, every nontrivial image line L has a *uniquely determined quark content* (cf. Remark 8.3.1), obtained by counting the highest power of each L_j or $\overline{L}_j$ in *any* of the summands in (8.29) that project onto L. (This accounts for the n flavors the L_j stand for as in Remark 8.3.6, the "heavy" part of the quark content being fixed throughout the discussion.)

In this way we have found mutually orthogonal lines in the SU(n) space of any SU(n) multiplet of hadrons, representing all individual (anti)baryons and those mesons which are not strictly neutral (see (vi) below). Each of these lines determines the quark content of the corresponding hadron, in the manner just described. Note that, by Remark

8.2.9, the distribution of quark contents in any given baryon multiplet with the SU(n) space $V_n^{2,1}$ does not depend on how $V_n^{2,1}$ is realized as an SU(V_n)-invariant subspace of $V_n V_n V_n$.

(vi) In contrast with baryons (§2.13), mesons existing in nature cannot be naturally divided into "mesons proper" and "antimesons", i.e., there is no conserved "meson number". Describing mesons as composite objects, we can therefore assume that in all tensor-product expressions quarks precede antiquarks. Thus, by switching factors, we identify the space $V_n \overline{V}_n = V_n \otimes \overline{V}_n$ in (iii)d with its conjugate $\overline{V}_n V_n$. In other words, we expect that the class of n^2 mesons in (iii)d is closed under antiparticle formation (§1.6) which, after all, transforms it into a class with essentially the same quark contents, quark composition, and the same "reducible SU(n) space" $V_n \overline{V}_n = \overline{V}_n V_n$. As a result, the meson forming the SU(n) singlet in (iii)d is strictly neutral, while *every* SU(n) *multiplet of* $n^2 - 1$ *mesons coincides with its own antimultiplet* (and, in fact, contains $n - 1$ strictly neutral particles, as explained below).

More precisely, we can define the *natural antiautomorphism* Φ of $V_n \overline{V}_n$ to be the switch map followed by the canonical antiisomorphism $V_n \leftrightarrow \overline{V}_n$ applied to either factor. We expect Φ to send lines in $V_n \overline{V}_n$ representing individual mesons onto those associated with their respective antiparticles, since this is the case for the lines $L_j \overline{L}_k$ with $j \neq k$ in (8.29), which do correspond to mesons (see (v)), and satisfy $\Phi(L_j \overline{L}_k) = L_k \overline{L}_j$ (while each L_j stands for a different quark flavor). On the other hand, $V_n \overline{V}_n = W + W^\perp$ with $W = \sum_j L_j \overline{L}_j$, $W^\perp = \sum_{j \neq k} L_j \overline{L}_k$, so that $W \subset \mathrm{Ker}(\Phi - \mathrm{Id})$. Thus, all mesons described by lines contained in W, i.e., orthogonal to all $L_j \overline{L}_k$ with $j \neq k$, are strictly neutral (§1.6), whether or not these lines are the $L_j \overline{L}_j$.

Note that, if $2 \leq n \leq 6$, the mesons represented by $L_j \overline{L}_k$ with $j \neq k$ are *not* strictly neutral, as for each of them one of the generalized charges I_3, S, C, b, t is nonzero (§8.1).

(vii) By (vi), the subspace $W = \sum_j L_j \overline{L}_j$ of $V_n \overline{V}_n = \mathrm{End}\, V_n$ stands for all strictly neutral mesons with a given quark composition, whose all constituent (anti)quarks belong to the n lightest flavors. Moreover, the direct summands $\mathbb{C} \cdot \mathrm{Id}_{V_n}$ and $W_0 = W \cap \mathfrak{sl}(V_n)$ of W (cf. (8.22)) represent the SU(n) singlet in (iii)d and, respectively, a part of the multiplet of $n^2 - 1$ mesons. To describe their member particles, it is necessary (see §8.3 or (b) of §1.9) to *select* complex lines K_α, $\alpha = 1, \ldots, n-1$, spanning W_0, and *mutually orthogonal*, according to the discussion following (6.26). (Thus, in contrast with (v), one cannot just use here projections of the n lines $L_j \overline{L}_j$ onto W_0, i.e., onto $\mathfrak{sl}(V_n) \subset V_n \overline{V}_n$.) It is possible to choose the K_α in various relatively natural ways, but, as a result, *a strictly neutral meson may in general be expected to have a mixed quark content* in the sense that its line K_α lies in the span of more than one $L_j \overline{L}_j$ (while each L_j stands for a different quark flavor, as in (8.27)). This obviously applies to the meson singlet described by $\mathbb{C}\cdot\mathrm{Id}_{V_n}$ as well.

Attempts to exhibit these mixed contents by selecting the K_α on the basis of naturality and simplicity always lead to results *depending on n* (cf. (viii)). This indicates that reality is more complicated and involves mixing among the quark-antiquark pairs $u\bar{u},\ldots,t\bar{t}$, as well as *mixing between specific multiplets* (see (viii)), the precise form of which is related to quark dynamics.

(viii) Every SU(n) multiplet of hadrons or quarks is contained in a unique SU($n+1$) multiplet. In fact, this is evident for quarks (§8.3). Thus, identifying the line summands L_j in (8.27) with quark flavors, we may regard V_n (along with its inner product) as a subspace of V_{n+1}. The summands of (8.19), (8.20) with $V = V_n$ then are contained in the corresponding summands for V_{n+1}, since the projection formulae in (i), (ii) of Remark 8.2.8 are dimension-independent. For the same reason, each line in these summands for V_n, standing for an individual baryon (see (v)) is one of the analogous lines for V_{n+1}, which establishes our claim for baryon multiplets of types (ii)c),d). On the other hand, $\mathfrak{sl}(V_n) \subset \mathfrak{sl}(V_{n+1})$ under the trivial-extension relation $\mathrm{End}\, V_n \subset \mathrm{End}\, V_{n+1}$ corresponding to the tensor-product inclusion $V_n\overline{V}_n \subset V_{n+1}\overline{V}_{n+1}$ (cf. (8.22)). Consequently, if the lines $K_\alpha \subset \mathfrak{sl}(V_n)$ representing the $n-1$ strictly neutral mesons contained in the SU(n) multiplet of (iii)d) are chosen so as to yield mixed quark contents that do not depend on n (see (vii)), these mesons (along with the other multiplet members, described by $L_j\overline{L}_k$ with $j \neq k$, cf. (v)) will explicitly become a part of an SU($n+1$) multiplet. Thus, the above assertion is valid for all hadrons, the cases of SU(n) singlets and n-tuplets being obvious.

Formally, however, the singlet in (iii)d) leads to some problems, since the $\mathbb{C}\cdot\mathrm{Id}$ summand of (8.22) with $V = V_n$ is contained in neither such should also have "correct" numbers of elements and assemblies of values for $Q, \ldots, t$, mentioned above. Moreover, if $n \geq 3$, any of these sets must be a union of $\mathrm{SU}(n-1)$ multiplets (cf. (viii)).

The resulting picture is consistent with experimental evidence leading, for $n = 2, 3, 4$, to an $\mathrm{SU}(n)$ multiplet pattern that fits the observed hadron spectrum quite well (even though many multiplets are incomplete, their missing members still awaiting discovery). The lightest $\mathrm{SU}(n)$ multiplets with $n = 2, 3$ are listed in the tables of §1.15 and in §8.6.

References: Lichtenberg 1978; Ryder 1986; Sudbery 1986.

8.5 Quarks and Isospin

The general description of $\mathrm{SU}(n)$ multiplets of hadrons, obtained in (ii), (iii) of §8.4 with the aid of the quark model, has some interesting consequences for isospin ($n = 2$). First of all, it implies that the only values $I \in \frac{1}{2}\mathbb{Z}_+$ actually assumed by the isospin (i.e., $\mathrm{SU}(2)$ spin) of hadrons, are $0, \frac{1}{2}, 1, \frac{3}{2}$ for (anti)baryons and $0, \frac{1}{2}, 1$ for mesons. In other words, by (2.13), the isospin symmetry may form meson singlets, doublets and triplets, as well as (anti)baryon singlets, doublets, triplets and quadruplets, while *no other hadron multiplets are possible*. This is consistent with the observed spectrum of hadrons (§1.15). As a result, we are dealing here with yet another fact of nature, which can be successfully explained by the quark model, but does not follow just from the hadron-based approach to isospin in §2.10 .

Furthermore, let $r \geq 0$ be the number of times the u or d flavor is represented among the constituent (anti)quarks of the given hadron, so that r is the same for all hadrons forming any given isospin multiplet, with $r \leq 3$ for (anti)baryons and $r \leq 2$ for mesons. According to (ii), (iii) of §8.4, $r - 2I$ is always a *nonnegative even integer*. This leads to the following conclusions about the strangeness S, charm C, bottomness b and topness t, each of which, like r, is constant on every isospin multiplet (Remark 8.1.3), and vanishes for the u and d quarks (§8.1):

(i) $S = C = b = t = 0$ for all (anti)baryon quadruplets and meson triplets, since $I = \frac{3}{2}$ implies $r = 3$, while $r = 2$ whenever $I = 1$.

(ii) For each (anti)baryon triplet and every meson doublet, one of S, C, b, t equals ± 1 and the other three are zero, as $I = 1$ yields $r = 2$, while $r = 1$ if $I = \frac{1}{2}$ and $r \leq 2$.

(iii) One of S, C, b, t must be nonzero for any baryon singlet, since r then equals 0 or 2.

(iv) Every meson triplet is its own antitriplet. In fact, by (vi) of §8.4, it contains a strictly neutral particle. (See also the beginning of §2.12.)

Assertions (i) – (iv) provide further examples of properties of hadrons, confirmed by experimental evidence, which do not follow from a theory treating hadrons as elementary (§2.10), yet are immediate consequences of the quark model. For other facts, correctly predicted by the quark model, see §6.2 and Remarks 8.1.2, 8.1.4.

References: Ryder 1986; Sudbery 1986.

8.6 Unitary Symmetry

Instead of SU(3) symmetry, SU(3) multiplets, SU(3) spin and SU(3) class, one often speaks of *unitary symmetry, supermultiplets, unitary spin* and *triality*, respectively.

The structure of supermultiplets of hadrons, predicted by the quark model, is obtained by setting $n = 3$ in (ii), (iii) of §8.4. Specifically, the *only possible* types of hadron supermultiplets are singlets, triplets and octets formed by (anti)baryons or mesons, as well as sextets and decuplets of (anti)baryons. The corresponding unitary spins $(s_1, s_2) \in \left(\frac{1}{3}\mathbb{Z}_+\right)^2$ are $(0,0)$ for singlets, $\left(\frac{1}{3}, 0\right), \left(0, \frac{1}{3}\right)$ for triplets, $\left(\frac{1}{3}, \frac{1}{3}\right)$ for octets, $\left(\frac{2}{3}, 0\right)$ for baryon sextets, $(1, 0)$ for baryon decuplets, and, respectively, $\left(0, \frac{2}{3}\right), (0, 1)$ for sextets and decuplets of antibaryons. Using the parameter r defined in (ii) of §8.4, one establishes as in §8.5 the following facts about the quantum numbers C, b, t (each of which is constant on any supermultiplet, cf. Remark 8.3.4):

(i) $C = b = t = 0$ for all hadron decuplets or octets, since their member particles contain u, d and s (anti)quarks only.

(ii) One of C, b, t is nonzero for each triplet or sextet (due to presence of quarks or antiquarks heavier than s). In fact, for sextets two of C, b, t vanish and the third equals ± 1.

(iii) Every meson octet is its own antioctet, and contains two strictly neutral particles (by (vi) of §8.4).

Physicists realized by early 1960s that hadrons are subject to a symmetry based on the group SU(3). However, it was only the quark model, developed by Gell-Mann and Zweig in 1964, that provided simple explanations for various properties of the SU(3) symmetry, such as (i), (ii) and (iii) above.

According to (viii) of §8.4, each supermultiplet is a union of isospin multiplets. For instance, the supermultiplets containing the lightest (anti)baryon multiplets (listed in Table 1.15.2) are the octets $N \cup \Lambda \cup \Sigma \cup \Xi$, $\overline{N} \cup \overline{\Lambda} \cup \overline{\Sigma} \cup \overline{\Xi}$ (with the *lighter* Λ and Σ) and the heavy-Λ singlets $\overline{\Lambda}, \Lambda$, while either of $\Delta \cup \Sigma$ and $\overline{\Delta} \cup \overline{\Sigma}$ (with the heavier Σ) is a part of a decuplet. As for the lightest mesons (Table 1.15.1), $\pi \cup K \cup \overline{K} \cup \eta$ and $\rho \cup \omega \cup K^* \cup \overline{K}^*$ are octets, η', ϕ are singlets, and $S \cup \delta$ is contained in an octet.

References: Ryder 1986; Lichtenberg 1978; Huang 1982; Sudbery 1986.

8.7 Quark Composition of Hadrons

The quark compositions (8.26) of hadrons are nothing else than the projection morphisms of the respective tensor-product bundles onto irreducible natural summands η (cf. §1.10, §1.13 and Remark 8.7.2 below). However, instead of trying to classify *all* such summands, it is useful to first eliminate some of them for the following reasons.

(i) Among the irreducible natural bundles over the given spacetime $(\mathcal{M}, g)$, only those listed in §1.16 and §1.17 actually represent matter particles (see §3.9).

(ii) According to (2.24), the baryon number B and orbital angular momentum ℓ of any hadron (see (8.26)) uniquely determine its parity ϵ (thus equal to $(-1)^\ell$ for baryons proper, $(-1)^{\ell+1}$ for antibaryons and mesons, cf. §8.1). Summands violating this "experimental" rule have to be discarded.

(iii) Multiple (i.e., repeated) summands, common in the case of baryons and antibaryons, count only once. See Remark 8.7.13.

As a consequence of (iii), the "physical" quark compositions are suitable *equivalence classes* of the morphisms (8.26), rather than these morphisms themselves (Remark 8.7.13) and so, by (i), (ii) and §§1.16, 1.17, they are completely determined by the spin s, orbital angular momentum ℓ, and baryon number B of the resulting hadron. Given ℓ and B, the values actually assumed by s form the set

$$\{\ell + N/2 - k : k \in \mathbb{Z}, \ 0 \leq k \leq \min(N, \ell + N/2)\}, \tag{8.30}$$

where $N = |B| + 2$ is the number of constituent (anti)quarks.

In other words, the possible values of the spin s for any prescribed orbital angular momentum ℓ are *all nonnegative numbers obtained as combinations with coefficients ± 1 of ℓ and the spins of the constituent quarks*. For details, see Remarks 8.7.10, 8.7.12.

Some of the projections (8.26) are explicitly described in Remarks 8.7.5, 8.7.9, 8.7.11, 8.7.12.

In the following remarks, σ stands for a fixed Dirac spinor bundle (§1.16, §6.3) over the given spacetime $(\mathcal{M}, g)$. We also write $1 = \mathcal{M} \times \mathbb{C}$ (cf. (5.2)), and identify the tangent bundle $T = T\mathcal{M}$ with its dual $T^* = T^*\mathcal{M}$ via g. As in §5.1, ordinary addition and multiplication is used to denote the direct sums and tensor products of vector bundles or (equivalence classes of) representations.

Remark 8.7.1. According to §1.18 and §3.7, natural bundles η over the given spacetime $(\mathcal{M}, g)$ are classified by equivalence classes of irreducible representations of the Lorentz group $\mathrm{O}^\uparrow(3,1)$, each of which is in turn characterized by the unordered pair of its spin invariants $s_1, s_2 \in \frac{1}{2}\mathbb{Z}_+$

and, if $s_1 = s_2$, also by its parity $\epsilon = \pm 1$. For the subsequent discussion, the equivalence class corresponding to such s_1, s_2 will be denoted $(s_1 + s_2)_\pm$ if $s_1 = s_2$ and $\epsilon = \pm 1$, $(s_1 + s_2)$ whenever $|s_1 - s_2| = \frac{1}{2}$, and $(\max(s_1, s_2), \min(s_1, s_2))$ in the case where $|s_1 - s_2| \geq 1$. Whenever the resulting symbols $(k)_+, (k)_-$ (or $(k)_\epsilon$ with $\epsilon = \pm 1$), $(s), (s_1, s_2)$ are used, we tacitly *assume* that $k \in \mathbb{Z}_+, s \in (\frac{1}{2}\mathbb{Z}_+) \smallsetminus \mathbb{Z}$ and $s_1, s_2 \in \frac{1}{2}\mathbb{Z}_+$ with $s_1 - s_2 \geq 1$.

According to §§1.16, 1.17 and 3.8, the natural bundles associated with $(k)_\epsilon$ (or, (s)) represent matter particles of spin k and parity ϵ (or, of spin s), while those corresponding to (s_1, s_2) are not matter-particle models.

Remark 8.7.2. Every finite dimensional complex representation of $\mathrm{O}^\uparrow(3,1)$ admits an irreducible decomposition (§3.4). If the summands are mutually nonequivalent, they can obviously be found just by exhibiting the corresponding projections, which are only required to be nonzero equivariant morphisms into irreducible representation spaces having the correct sum of dimensions. By Remark 8.7.1, the same is true for natural bundles over $(\mathcal{M}, g)$.

Remark 8.7.3. The irreducible natural summands of $\sigma\overline{\sigma}$ are the complexifications of

$$1, \quad \Lambda^4 T, \quad T, \quad (\Lambda^4 T) T, \quad \Lambda^2 T. \tag{8.31}$$

This is clear from Remark 8.7.2 applied to the projection morphisms sending $\psi\phi \in \sigma_x\sigma_x$, $x \in \mathcal{M}$, onto the scalar $\omega(\psi, \phi)$, pseudoscalar $\omega(I\psi, \phi)$, vector $p \mapsto \omega(\psi, \gamma(p)\phi)$, pseudovector $p \mapsto \omega(I\psi, \gamma(p)\phi)$, and bivector $p \wedge q \mapsto \omega(\gamma(p)\psi, \gamma(q)\phi) - \omega(\gamma(q)\psi, \gamma(p)\phi)$, respectively, where $p, q \in T_x^*\mathcal{M}$, while ω, γ, I are defined as in §1.3 or §6.3, and then extended complex (bi)linearly to σ. See also Remark 8.7.4.

Remark 8.7.4. For a spacetime $(\mathcal{M}, g)$ and $x \in \mathcal{M}$, any $\psi \in \Lambda^4 T_x\mathcal{M}$ can be regarded as a *pseudoscalar* in $T_x\mathcal{M}$, i.e., a number defined up to a sign, in such a way that the choices of its sign are identified with orientations of $T_x\mathcal{M}$. In fact, the number $\pm a$ corresponding to ψ is given by $\psi = a e_0 \wedge \ldots \wedge e_3$, where the e_μ form any g-orthonormal basis of $T_x\mathcal{M}$. Similarly, a *pseudovector* in $T_x\mathcal{M}$ is a vector determined only up to an orientation-dependent sign. Pseudoscalars and pseudovectors appear in Remark 8.7.3 since the sign of I_x changes when the orientation of $T_x\mathcal{M}$ is reversed (see §1.3 or §6.3).

Remark 8.7.5. In view of the natural isormorphism $\overline{\sigma} = \sigma$ (cf. (6.6)), Remark 8.7.3 provides an irreducible decomposition of $\sigma\overline{\sigma}$ as well. However, only two of the five summands corresponding to (8.31), namely, the complexifications of $\Lambda^4 T$ and T, give rise to quark compositions (8.26) of mesons with $\ell = 0$, since the other three are eliminated by (i), (ii) above (cf. §§1.16,

1.17). The resulting morphisms (8.26) from $\sigma\bar{\sigma}$ into complex pseudoscalars and vectors are defined by $\psi\phi \mapsto \omega(I\psi, \bar{\phi})$ and $\psi\phi \mapsto \omega(\psi, \gamma(\cdot)\bar{\phi})$, respectively.

Remark 8.7.6.   On the Lie algebra level, an irreducible representation of $O^{\uparrow}(3,1)$ has at most two irreducible summands (§3.7). Restricted to the isotropy algebra $\mathfrak{so}(3) \subset \mathfrak{so}(3,1)$ of a timelike vector (§3.4), such a representation, described as in Remark 8.7.1, has irreducible summands with spins $k, k-1, \ldots, 0$ for $(k)_{\pm}$, each occurring once, $s, s-1, \ldots, \frac{1}{2}$ for (s) and $s_1 + s_2, s_1 + s_2 - 1, \ldots, s_1 - s_2$ for (s_1, s_2), each repeated twice (see §3.4). Thus, for any representation φ of $O^{\uparrow}(3,1)$, the numbers of irreducible summands of φ (each counted with its multiplicity) that are of type $(k)_{\pm}$ or (s), i.e., actually correspond to matter particles (Remark 8.7.1) are, respectively, n_0 and $\frac{1}{2}n_{1/2}$, where n_s is the number of $\mathfrak{so}(3)$-irreducible summands of φ with spin s. In view of (3.1), these numbers are easy to evaluate for tensor products of irreducible representations such as those involved in (8.26).

Remark 8.7.7.   According to §3.8, a Dirac spinor bundle σ and the natural bundles over $(\mathcal{M}, g)$ obtained by complexifying $1, \Lambda^4 T, T, (\Lambda^4 T)T, S_0^\ell T$ and $\Lambda^2 T$ correspond, in the notation of Remark 8.7.1, to the irreducible representations $\left(\frac{1}{2}\right), (0)_+, (0)_-, (1)_-, (1)_+, (\ell)_\epsilon$ with $\epsilon = (-1)^\ell$, and $(1,0)$, respectively.

Remark 8.7.8.   By Remarks 8.7.3 and 8.7.7,

$$\left(\frac{1}{2}\right)\left(\frac{1}{2}\right) = (0)_+ + (0)_- + (1)_+ + (1)_- + (1,0). \tag{8.32}$$

On the other hand, we have

$$\begin{aligned} (k)_\delta (0)_\epsilon &= (k)_{\delta\epsilon}, \quad \epsilon, \delta \in \{1, -1\}, \\ (s)(0)_\pm &= (s), \quad (s_1, s_2)(0)_\pm = (s_1, s_2), \end{aligned} \tag{8.33}$$

as one sees using $\mathfrak{so}(3)$-irreducible decompositions (Remark 8.7.6) and the fact that multiplication by $(0)_\pm$ does not affect $O^{\uparrow}(3,1)$-irreducibility of the representation.

Remark 8.7.9.   For any integer $\ell \geq 1$ and $\epsilon, \delta \in \{1, 1\}$, $(\ell)_\epsilon(1)_\delta \sim (\ell - 1)_{\epsilon\delta} + (\ell+1)_{\epsilon\delta}$ and $(\ell)_\epsilon(1,0) \sim (\ell)_\epsilon$, where $\sim$ means that the left-hand side equals the right-hand side plus (zero or more) summands of the form (s_1, s_2) (notation of Remark 8.7.1). In fact, one easily verifies as in Remark 8.7.6 that the above relations render the correct number of "physical" summands (of type $(k)_\pm$) in either tensor product. These summands can in turn be identified by assuming that $\epsilon = (-1)^\ell, \delta = -1$ (which is allowed in view of (8.33)) and exhibiting the corresponding (real) projections $(S_0^\ell T)T \to S_0^{\ell-1} T$, $(S_0^\ell T)T \to S_0^{\ell+1} T$ and $(S_0^\ell T)\Lambda^2 T \to S_0^\ell T$, where $T = T^*\mathcal{M}$ for a fixed spacetime $(\mathcal{M}, g)$. Specifically, the first projection is the contraction $(v_1 \odot \ldots \odot v_\ell) \otimes v \mapsto \sum_{j=1}^{\ell} g(v, v_j) v_1 \odot \ldots \odot v_{j-1} \odot v_{j+1} \odot \ldots \odot v_\ell$, the second is obtained by projecting onto the kernel of the first and then applying the symmetric multiplicaton $(v_1 \odot \ldots \odot v_\ell) \otimes v \mapsto v_1 \odot \ldots \odot v_\ell \odot v$, while the third is just the action, at any $x \in \mathcal{M}$, of the Lie algebra $\mathfrak{so}(T_x\mathcal{M}) \cong \Lambda^2 T_x \mathcal{M}$ on $S_0^\ell T_x \mathcal{M}$, induced by the obvious group action of $O(T_x\mathcal{M}; g_x) \cong O(3,1)$.

Remark 8.7.10. By (8.32), (8.33) and Remark 8.7.9, for any integer $\ell \geq 1$,

$$(\ell)_\pm \left(\frac{1}{2}\right)\left(\frac{1}{2}\right) \sim (\ell)_\pm + \sum_{\epsilon \in \{1, -1\}} [(\ell)_\epsilon + (\ell - 1)_\epsilon + (\ell + 1)_\epsilon].$$

These summands correspond to the possible projections $(S_0^\ell T^*) \sigma \bar{\sigma} \to \eta$ in (8.26), except for those eliminated by (i) above. However, condition (ii) then imposes further restrictions, so that with $\epsilon = (-1)^\ell$, the only "genuinely physical" summands left in $(\ell)_\epsilon \left(\frac{1}{2}\right)\left(\frac{1}{2}\right)$ are $(\ell)_{-\epsilon}, (\ell - 1)_{-\epsilon}$ and $(\ell + 1)_{-\epsilon}$, each occurring only once.

Remark 8.7.11. We have $\left(\frac{1}{2}\right)(1)_\pm = \left(\frac{1}{2}\right) + \left(\frac{3}{2}\right)$ and $\left(\frac{1}{2}\right)(1, 0) = 2\left(\frac{1}{2}\right) + 2\left(\frac{3}{2}, 0\right)$ (notation of Remarks 8.7.1, 8.7.8, with integral coefficients indicating multiplicities). In fact, the image and kernel of the Clifford multiplication $T^* \sigma \to \sigma$ correspond to the first pair of summands (Example 3.8.2). On the other hand, at any point x of the spacetime $(\mathcal{M}, g)$, the double-valued action of $O^\uparrow(T_x\mathcal{M}, g_x)$ on the spinor space σ_x induces a Lie algebra action of $\mathfrak{so}(T_x\mathcal{M}) \cong \Lambda^2 T_x \mathcal{M}$ on σ_x, which gives rise to a projection $\mathrm{pr} : \sigma \cdot \Lambda^2 T \to \sigma$. Another such projection is $\mathrm{pr} \circ (\mathrm{Id}_\sigma \otimes *)$, where $* : \Lambda^2 T \to \Lambda^2 T$ is the Hodge star. This shows that $\left(\frac{1}{2}\right)(1, 0)$ has two $\left(\frac{1}{2}\right)$ summands (cf. Remark 8.7.7) and its remaining summands are easily determined using $\mathfrak{so}(3)$-irreducible decompositions as in Remark 8.7.6.

Remark 8.7.12. By (8.32), (8.33) and Remark 8.7.10,

$$\left(\frac{1}{2}\right)\left(\frac{1}{2}\right)\left(\frac{1}{2}\right) = 6\left(\frac{1}{2}\right) + 2\left(\frac{3}{2}\right) + 2\left(\frac{3}{2}, 0\right).$$

From now on, let ℓ be a fixed positive integer. Then $(\ell)_\pm \left(\frac{1}{2}\right) = \left(\ell + \frac{1}{2}\right) + \left(\ell - \frac{1}{2}\right)$, which one verifies just as in Remark 8.7.10 (where $\ell = 1$), using the Clifford multiplication involving σ and one of the T^* factors in $S_0^\ell T^*$. Moreover, if $\ell \geq 1$, $\left(\ell + \frac{1}{2}\right)(1, 0) \sim 4\left(\ell + \frac{1}{2}\right)$ (notation of Remark 8.7.9), for similar reasons as in Remark 8.7.9 with $\ell = 0$ (with two extra summands of type $\left(\ell + \frac{1}{2}\right)$ coming from the action of $\mathfrak{so}(T_x\mathcal{M}) \cong \Lambda^2 T_x \mathcal{M}$ on $S_0^\ell T_x \mathcal{M}$). Finally, $\left(\ell + \frac{1}{2}\right)(1)_\pm \sim \left(\ell - \frac{1}{2}\right) + \left(\ell + \frac{1}{2}\right) + \left(\ell + \frac{3}{2}\right)$, which can be obtained as in Remark 8.7.9 using the contraction, Clifford-product, and symmetric-product projections. Applying the above relations (with ℓ replaced by $\ell - 1$ if necessary) to (8.32) and using (8.33), we obtain

$$(\ell)_{\pm} \left(\frac{1}{2}\right) \left(\frac{1}{2}\right) \left(\frac{1}{2}\right) \sim \sum_{j=0}^{\min(3,\ell+3/2)} k_j \left(\ell + \frac{3}{2} - j\right)$$

with some positive integers k_j (that may also depend on ℓ, at least if ℓ is small). Along with Remark 8.7.10, this proves the classification (8.30) of physical quark compositions of hadrons.

Remark 8.7.13. By Remarks 8.7.10, 8.7.12, the summands η of the tensor-product bundles in (8.26), which satisfy the "selection rules" (i), (ii) at the beginning of this section, are all simple (pairwise nonequivalent) for mesons, and multiple for (anti)baryons. However, the observed baryon spectrum indicates that repeated summands correspond to a *single* particle species in nature. Thus, two projections $(S_0^\ell T^*)\sigma^3 \rightarrow \eta$ may be called *physically equivalent* if one is the composite of the other with a natural bundle automorphism of $(S_0^\ell T^*)\sigma^3$. It is these equivalence classes, rather than the projections themselves, that are of real, physical significance.

On the other hand, each of the meson projection morphisms $S_0^\ell \sigma \bar{\sigma} \rightarrow \eta$ is unique up to a constant factor. (According to (b) of §1.2, such factors are physically irrelevant.)

Reference: Ryder 1986.

9 Hadrons and the Weak Interaction*

The main topic discussed in this chapter is the geometric structure of the *Glashow-Illiopoulos-Maiani model,* which is an extension of the electroweak theory to quarks, and through them, to hadrons. Additional complexities involved in such an extension include the phenomenon of *weak mixing* of quark flavors, the extent of which is characterized by an angular parameter introduced by Cabibbo and later generalized by Kobayashi and Maskawa.

9.1 Natural Subspaces of Direct Sums

Let $V \oplus \ldots \oplus V$ be the direct sum of n copies, $n \geq 2$, of a finite dimensional complex vector space V, possibly endowed with some additional geometric with $\theta_C = \frac{\pi}{2}$ if $a_1 = 0$. Thus,

$$\cos\theta_C = |a_1|, \quad \sin\theta_C = |a_2|, \tag{9.4}$$

since we always assume that $|a_1|^2 + |a_2|^2 = 1$.

The geometric significance of the Cabibbo angle and its generalizations to $n > 2$ are discussed below.

Remark 9.1.1. Each of the subspaces (9.1) is invariant under the diagonal action of the full automorphism group $GL(V) = \mathrm{Aut}\,V$ on $V \oplus \ldots \oplus V$. Conversely, let $G \subset GL(V)$ be a subgroup (such as the group of all automorphisms preserving the geometry of V), which acts on V irreducibly in the sense of having no nontrivial proper invariant subspaces. All G-invariant (i.e., natural) nontrivial subspaces V' of $V \oplus \ldots \oplus V$ which are either G-irreducible, or satisfy $\dim V' \leq \dim V$, then must be of the form (9.1). This is immediate from Schur's lemma applied to transitions between those projections $V' \to V$ onto summands of $V \oplus \ldots \oplus V$ which are nonzero, and hence isomorphic.

Remark 9.1.2. Generalizing the Cabibbo angle to the case of any $n \geq 2$, one associates with an orthogonal n-tuple in $V \oplus \ldots \oplus V$, represented as above by a matrix $a = [a_{jk}] \in U(n)$, the *angular parameters* $\theta_1, \ldots, \theta_{2n-3} \in \left[0, \frac{\pi}{2}\right]$ with

$$|a_{11}| = c_1, \quad |a_{1n}| = \prod_{k=1}^{n-1} s_{n+k-1}, \quad |a_{n1}| = \prod_{k=1}^{n-1} s_k,$$

$$|a_{1j}| = c_{n+j-1} \prod_{k=1}^{j-1} s_{n+k-1}, \quad |a_{j1}| = c_j \prod_{k=1}^{j-1} s_k \quad \text{if} \quad 1 < j < n, \tag{9.5}$$

where

$$c_\ell = \cos\theta_\ell, \quad s_\ell = \sin\theta_\ell \tag{9.6}$$

(cf. (9.4)). In other words, $(\theta_1, \theta_n, \theta_{n+1}, \ldots, \theta_{2n-3})$ and $(\theta_1, \theta_2, \ldots, \theta_{n-1})$ are the spherical coordinates of the unit vectors $(|a_{11}|, \ldots, |a_{1n}|)$ and $(|a_{11}|, \ldots, |a_{n1}|)$ of $\mathbb{R}^n$. Thus, in contrast with the Cabibbo angle $\theta_C = \theta_1$ for $n = 2$, for some orthogonal n-tuples the angular parameters θ_ℓ with $n \geq 3$ may fail to be uniquely determined by (9.5). See also Remarks 9.1.4 – 9.1.6.

Remark 9.1.3. The orthogonal n-tuple in $V \oplus \ldots \oplus V$ determined by a matrix $a \in U(n)$ remains unchanged if a row of a is multiplied by $z \in \mathbb{C}$ with $|z| = 1$, i.e., if a is replaced by qa for any q in the maximal torus T^n consisting of all diagonal matrices in $U(n)$. This establishes a bijective correspondence between the set of all orthogonal n-tuples in $V \oplus \ldots \oplus V$ and the coset space (flag manifold) $U(n)/T^n$.

Remark 9.1.4. It is natural to call two orthogonal n-tuples in $V \oplus \ldots \oplus V$ *geometrically equivalent* if one is transformed into the other by an automorphism Φ of $V \oplus \ldots \oplus V$ such that all summands are contained in eigenspaces of Φ with eigenvalues of modulus 1. In terms of unitary matrices $a \in \mathrm{U}(n)$ parametrizing the orthogonal n-tuples, geometric equivalence can be realized by multiplying each column of a by the corresponding eigenvalue, i.e., by replacing a with aq, where $q \in \mathrm{T}^n$. Thus, by Remark 9.1.3, the set $\mathcal{K}_n$ of geometric equivalence classes of orthogonal n-tuples in $V \oplus \ldots \oplus V$ is naturally identified with the orbit space $\mathrm{U}(n)/\mathrm{T}^n \times \mathrm{T}^n$, where the T^n factors act by left and right multiplications, respectively. The angular parameters θ_ℓ (whenever defined, see Remark 9.1.2), including the Cabibbo angle θ_C when $n = 2$, are obviously not affected by geometric equivalence, i.e., remain unchanged after a row or column of the matrix $a \in \mathrm{U}(n)$ has been multiplied by a unit complex number (cf. (9.5)). Thus, θ_ℓ may be regarded as functions on (a subset of) $\mathcal{K}_n$, valued in $\left[0, \frac{\pi}{2}\right]$. If $n = 2$, the Cabibbo angle $\theta_C = \theta_1 : \mathcal{K}_2 \to \left[0, \frac{\pi}{2}\right]$ is easily seen to be bijective, i.e., an orthogonal pair is determined up to geometric equivalence by the value of θ_C. This is not the case for $n \geq 3$, where besides the θ_ℓ, other invariants must be used to characterize elements of $\mathcal{K}_n$ (see Remark 9.1.5).

Remark 9.1.5. Let $\mathcal{K}'_n$ be the open dense subset of $\mathcal{K}_n = \mathrm{U}(n)/\mathrm{T}^n \times \mathrm{T}^n$ (Remark 9.1.4) determined by the set $\mathrm{U}'_n \subset \mathrm{U}(n)$ of all unitary matrices a with

$$a_{j1} \neq 0 \neq a_{1j} \quad \text{if } 2 \leq j \leq n. \tag{9.7}$$

We have a natural bijective mapping

$$(\theta_1, \ldots, \theta_{2n-3}, u) : \mathcal{K}'_n \to (0, \pi/2) \times (0, \pi/2)^{2n-4} \times \mathrm{U}(n-2) \tag{9.8}$$

where the θ_ℓ are determined (uniquely) by (9.5), (9.6) (cf. Remark 9.1.4), while $u : \mathcal{K}'_n \to \mathrm{U}(n-2)$ is defined as follows. Let $a \in \mathrm{U}'_n$ represent a given element of $\mathcal{K}'_n$. We may assume, multiplying the rows and columns of a by suitable factors of modulus 1 (cf. Remarks 9.1.3, 9.1.4) that

$$a_{j1} = |a_{j1}|, \quad a_{1k} = -|a_{1k}| \quad \text{if } 1 \leq j \leq n \text{ and } 2 \leq k \leq n. \tag{9.9}$$

A matrix $a \in \mathrm{U}'_n$ with these properties is obviously unique in its geometric equivalence class $a(\mathrm{mod}(\mathrm{T}^n \times \mathrm{T}^n)) \in \mathcal{K}'_n$, and so we may describe u as a function of a. Let $b_1, \ldots, b_n \in \mathbb{C}^{n-1}$ be the rows of the matrix b of rank $n - 1$ obtained from a by deleting the first column. As $\sum_{j=1}^n \bar{a}_{j1} b_j = 0$ and $a_{n1} \neq 0$, the vectors $b_1, \ldots, b_{n-1}$ form a basis of $\mathbb{C}^{n-1}$. The vector $v = b_1/|b_1|$ satisfies $v = yv_0 + zw_0$ for some $y, z \in \mathbb{C}$, where $v_0 = (1, 0, \ldots, 0) \in \mathbb{C}^{n-1}$ and $\{v_0, w_0\}$ is the pair of vectors in $\mathbb{C}^{n-1}$ obtained by orthonormalizing $\{v_0, b_1\}$. Thus, v and $w = -\bar{z}v_0 + \bar{y}w_0$ form an orthonormal basis of $W = \mathrm{Span}\{v_0, b_1\} \subset \mathbb{C}^{n-1}$. Let F be the linear isometry of $\mathbb{C}^{n-1}$ with $Fv = v_0$, $Fw = w_0$ and $F = \mathrm{Id}$ on $W^\perp$. Orthonormalizing the basis $Fb_1, \ldots, Fb_{n-1}$

of $\mathbb{C}^{n-1}$, we obtain a system $(1, 0, \ldots, 0)$, $(0, u_2), \ldots, (0, u_{n-1})$, where the vectors $u_2, \ldots, u_{n-1} \in \mathbb{C}^{n-2}$ are orthonormal, i.e., form the rows of the required matrix $u = u(a) \in \mathrm{U}(n-2)$. It is now easy to verify that (9.8) is bijective.

Remark 9.1.6. If $n = 2$, $\mathrm{U}(n-2)$ is trivial and so the parameter u in (9.8) is irrelevant (cf. Remark 9.1.4). On the other hand, for $n = 3$, $\theta_1, \theta_2, \theta_3$ and u in (9.8) are known as the *Kobayashi-Maskawa parameters* (associated with a given 3×3 unitary matrix $a \in \mathrm{U}'_3$). One usually calls $u = u(a) \in \mathrm{U}(1) = \mathrm{S}^1 \subset \mathbb{C}$ the *phase factor* of a, and writes $u = e^{i\delta}$ with $\delta \in \mathbb{R}$. In terms of these parameters, the unique matrix satisfying (9.9) and geometrically equivalent to the given $a \in \mathrm{U}'_3$, is easily verified to be the *Kobayashi-Maskawa matrix*

$$a_{\mathrm{KM}} = \begin{bmatrix} c_1 & -s_1 c_3 & -s_1 s_3 \\ s_1 c_2 & c_1 c_2 c_3 - s_2 s_3 u & c_1 c_2 s_3 + s_2 c_3 u \\ s_1 s_2 & c_2 s_2 c_3 + c_2 s_3 u & c_1 s_2 s_3 - c_2 c_3 u \end{bmatrix} \tag{9.10}$$

with $c_\ell = \cos\theta_\ell$, $s_\ell = \sin\theta_\ell$.

9.2 Quark Mixing and the Cabibbo Angle

A fundamental difference between the lepton generations (6.20) on the one hand, and quarks along with their generations (6.1) on the other, lies in the fact that the values of universally conserved particle invariants uniquely determine each (anti)lepton species, yet fail to do so for individual quark flavors, or even quark generations. More precisely, the invariants in question $(Q, B, L_e, L_\mu, L_\tau$ and, in a sense, the spin s, along with their functions such as L or $(-1)^{2s}$, cf. §2.14) can only be used to characterize the *quark triples* $\{\mathrm{u, c, t}\}$, $\{\mathrm{d, s, b}\}$ and the corresponding "antitriples" (which transcend the generation boundaries), but are incapable of distinguishing among the (anti)quark flavors forming a triple (see §8.1).

As a consequence, it is in general impossible to predict the identity of quarks emerging from weak interaction processes, where invariants other then those just listed are not always conserved. One can, however, make reasonable predictions about quark triples and, within each triple, about the statistical distribution of individual quark flavors (i.e., the probabilities with which they will appear in repeated experiments). The objects produced in such a process may therefore turn out to be not ordinary quark flavors but rather *mixtures* of these with some weights (coefficients) reflecting the above probabilities.

Mixtures of particles are, formally, a by-product of generalized-particle formation involving relatively similar species (§1.9). Thus, we have to decide which of the above quark triples, or their two-element subsets, are physically most suitable to be regarded as particle generalizations. The example of $SU(n)$ symmetry in §8.3 indicates that the masses of the quarks forming a physically meaningful generalized particle should be "close", i.e., not separated by other quark masses (without counting antiquarks), which applies to hadrons as well (see Remark 8.3.5). By (6.2), this requirement reduces our choice just to one possible generalization, namely the pair $\{d, s\}$ (along with its "antipair" $\{\bar{d}, \bar{s}\}$). Consequently, a quark-like object with the electric charge of $-\frac{1}{3}$ in proton charge units, involved in weak interactions, has to be *a mixture of the* d *and* s *quarks*, unless it is massive enough to be the b quark. (About the possibility of mixing d, s *and* b, see §9.3.)

Let σ be the fixed Dirac spinor bundle over the spacetime $(\mathcal{M}, g)$, representing all quark flavors (§6.2). A generalization (as in §1.8) of the d and s quarks then lives in the bundle $\sigma \oplus \sigma = \sigma_d \oplus \sigma_s$, the summands $\sigma_d = \sigma \oplus \{0\}$, $\sigma_s = \{0\} \oplus \sigma$ of which correspond to the d and s quark flavors. A quark-like object with the properties just stated then is described by a bundle $\sigma_{d'}$, isomorphic to σ, which is a summand in *another* decomposition of $\sigma_d \oplus \sigma_s$ (see §1.9). It is also natural to expect (see the discussion following (6.26)) that the other summand, denoted $\sigma_{s'}$, then is the orthogonal complement of $\sigma_{d'}$ relative to the indefinite fibre metric in $\sigma \oplus \sigma$ induced by (6.7). Similar reasons of naturality indicate that the fibres of $\sigma_{d'}$ and $\sigma_{s'}$ at any $x \in \mathcal{M}$ should be of the form (9.2), where V is the fibre of σ and a_1, a_2 may be chosen constant (independent of x) due to the obvious requirement that $\sigma_{d'}$ (and $\sigma_{s'}$) be invariant under the spinor connection extended to $\sigma \oplus \sigma$. The constants a_1, a_2 give rise to the *(physical) Cabbibo angle* θ_C, characterized by (9.3) and describing how the d and s quarks are affected by the weak interaction. More precisely, one thinks of $\sigma_{d'}$ and $\sigma_{s'}$ as representing two mutually "complementary" mixtures d', s' of d and s.

As mentioned before, θ_C can be interpreted statistically, which leads to predictions verifiable by experiment. As a result, one obtains the approximate value

$$\sin \theta_C = 0.231 \pm 0.003 \qquad (9.11)$$

characterizing the Cabibbo angle in nature.

References: Ryder 1986; Leite Lopes 1981; Mohapatra 1986; Huang 1982.

9.3 The Kobayashi-Maskawa Parameters

The fact that the pair $\{d, s\}$ is the most natural choice of a quark generalization pertinent to the weak interaction indicates that, for the quark triples introduced in §9.2, the members of $\{d, s, b\}$ are much more closely related than those of $\{u, c, t\}$. (The same applies to their antitriples.) Relaxing the previously used mass-range requirements, we can therefore try to generalize the symmetry between d and s, obtaining a (weaker) symmetry involving d, s and b. In other words, $\{d, s, b\}$ may be thought of as a new, less restrictive kind of a generalized particle, despite the mass separation by the c quark in (6.2). Starting form this idea, Kobayashi and Maskawa developed in 1973 a generalization of Cabibbo's theory. According to their model, the d, s and b quarks enter weak-interaction processes in the form of mixtures d', s', b', represented by a natural orthogonal decomposition of $\sigma \oplus \sigma \oplus \sigma = \sigma_d \oplus \sigma_s \oplus \sigma_b$ with some new summands $\sigma_{d'}, \sigma_{s'}, \sigma_{b'}$. In full analogy with the case of d, s mixing (§9.2), the fibres of these summands are expected to form, at each point $x \in \mathcal{M}$, an orthogonal triple in $\sigma_x \oplus \sigma_x \oplus \sigma_x$ (§9.1) that may be described by a constant (independent of x) matrix $a \in U(3)$. By Remark 9.1.5, it is plausible to assume (9.9) with $n = 3$. The matrix a then gives rise to the (physical) Kobayashi-Maskawa parameters $\theta_1, \theta_2, \theta_3, u$, consisting of the three angles in $[0, \frac{\pi}{2}]$ and a phase factor $u = e^{i\delta}$, $\delta \in \mathbb{R}$ (see Remark 9.1.6). These parameters characterize the geometric equivalence class of a (Remark 9.1.4), i.e., the geometric configuration of $\sigma_{d'}, \sigma_{s'}, \sigma_{b'}$ within $\sigma_d \oplus \sigma_s \oplus \sigma_b$, usually described by the Kobayashi-Maskawa matrix (9.10). Comparing the predictions of the Kobayashi-Maskawa model with experimental data, one obtains the estimates

$$0.216 < \sin\theta_1 < 0.238,$$
$$0.05 < \sin\theta_2 < 0.14, \qquad\qquad (9.12)$$
$$0.02 < \sin\theta_3 < 0.06$$

for the values of $\theta_1, \theta_2, \theta_3$ in nature.

The Cabibbo model may be thought of as a simplified, approximate version of the Kobayashi-Maskawa theory. More precisely, the former is the limiting case of the latter as $b' \to b$, i.e., when the b mixing vanishes. In fact, we then have $\sigma_{d'}, \sigma_{s'} \subset \sigma_d \oplus \sigma_s$, as $\sigma_{b'} = \sigma_b$. Moreover, the matrix $a \in U(3)$, characterizing the triple $\sigma_{d'}, \sigma_{s'}, \sigma_{b'}$ at each point $x \in \mathcal{M}$, then satisfies $|a_{13}| + |a_{23}| + |a_{31}| + |a_{32}| \to 0$ and $|a_{33}| \to 1$, so that, choosing $a = a_{KM}$ with (9.10), we obtain the following limit relations for the Kobayashi-Maskawa parameters: $\theta_2 \to 0$, $\theta_3 \to 0$ and $\theta_1 \to \theta_C$, where θ_C is the Cabibbo angle of the "limit pair" $\sigma_{d'}, \sigma_{s'}$ in $\sigma \oplus \sigma$ (cf. (9.5)). The good agreement between these limits with θ_C given by (9.11), and the experimental values (9.12) shows that Cabibbo's theory provides a reasonably accurate approximation to the Kobayashi-Maskawa model.

References: Leite Lopes 1981; Mohapatra 1986;

9.4 Extension of the Electroweak Model to Quarks

It might appear that, to extend the electroweak model to quarks, one should proceed just as in §6.5, except for replacing each of the lepton generations (6.20) by the corresponding basic-fermion generation in (7.2) and ignoring the effects of the strong interaction. However, this approach fails to account for the fact that the d and s quarks emerge from weak-interaction processes not in their pure form, but rather as mixtures d′, s′, the degree of mixing being characterized by the Cabibbo angle θ_C (§9.2). Therefore, the original generations (7.2) must be replaced by

$$(e, \nu_e, u, d'), \quad (\mu, \nu_\mu, c, s'), \quad (\tau, \nu_\tau, t, b'), \tag{9.13}$$

where b′ = b unless we use, instead of Cabibbo's theory and θ_C, the more general Kobayashi-Maskawa model and the parameters θ_1, θ_2, θ_3, u (§9.3).

We have thus arrived at the underlying idea of the *electroweak model for leptons and quarks*, developed by Glashow, Illiopoulos and Maiani in 1970. To describe the geometry of this model, we simply follow the discussion in §6.5 as closely as possible, modifying it as needed to accomodate (9.13) and the fractional electric charges of the quarks. Thus, the model accounts for each of the generations (9.13) separately and, for definiteness, we choose the generation we deal with to be (e, ν_e, u, d'). The electroweak-interaction bundle ι, with the structure group U(2), is the same as in §6.5, and a fixed Dirac spinor bundle σ over the spacetime $(\mathcal{M}, g)$ again represents the whole generation, and is referred to as the *generic-fermion bundle* (cf. also §7.1).

Inspired by (6.22) – (6.25), we interpret the Hermitian line bundle $\lambda = \Lambda^2 \iota$ as the *electromagnetism bundle* corresponding to the electron charge. Since the quark charges are fractional (§6.2), it is necessary to assume, as in §7.1, that we are also given a *cubic root* $\lambda^{1/3}$ of λ, i.e., a Hermitian line bundle χ along with a fixed vector-bundle isometry $\chi \otimes \chi \otimes \chi \to \lambda$ (cf. Remark 4.13.1). The system $(\iota, \lambda^{1/3})$ with the geometry just described can be interpreted as the vector bundle $\iota \oplus \lambda^{-1/3}$ endowed with a suitable G-structure (see Remark 9.4.1 below), so that we are still within the general framework of Yang-Mills theory (§5.4).

Using the notation of (4.29) (extended to rational numbers k of denominator 3) and (5.1), we may define

$$\alpha = \iota \sigma_L + (\Lambda^2 \iota)\sigma_R + \lambda^{-2/3} \iota \sigma \tag{9.14}$$

with σ_L, σ_R as in (6.4), to be the *interacting-fermion bundle*. Our selection of formula (9.14) among the options offered by (5.12) and leading (via symmetry breaking as in §6.5) to the correct description (9.15) of the generation (e, ν_e, u, d'), is motivated by its simplicity. It is *not*, however, the choice of α usually made by physicists (see Remark 9.4.3).

The spontaneous symmetry breaking in the electroweak model, already described in §6.5, consists in choosing a C^∞ section ϕ of ι with a constant length $|\phi| > 0$. As in (6.23), this gives rise to the decomposition $\iota = \mathbb{C}\phi + \phi^\perp = 1 + \lambda$ (notation (of (5.2)), so that (9.14) with (6.4) becomes

$$\alpha = \sigma_{\mathsf{L}} + \lambda\sigma + \lambda^{-2/3}\sigma + \lambda^{1/3}\sigma. \tag{9.15}$$

The four summands in (9.15) correctly represent the fermions ν_e, e, u, d' (§§1.16, 1.17, 6.2, 6.3) along with their electric charges $0, -1, \frac{2}{3}, -\frac{1}{3}$ in proton charge units (§6.5, §6.2 and Remark 4.13.1).

A similar description applies to the other generations in (9.13), while a model simultaneously dealing with two or more generations can be obtained by a direct-sum construction as in (v) of §6.6.

Electroweak-interaction carriers are, in this case, again represented by the affine bundle $\mathcal{C}(\iota)$ the sections of which are U(2) connections in ι, since such connections are automatically compatible with the choice of $\lambda^{1/3}$ (see Remark 9.4.2). Thus, the symmetry breaking leads to the same interaction-carrier particles as those obtained from (6.30): the photon γ, and the massive weak bosons W^-, W^+, Z^0.

Remark 9.4.1.   The bundles $\iota, \lambda = \Lambda^2\iota$ and $\chi = \lambda^{1/3}$ as described above, along with their geometries, can obviously be replaced by a complex vector bundle ζ of fibre dimension 3 over $(\mathcal{M}, g)$ endowed with a Hermitian fibre metric $\langle , \rangle$, a fixed orthogonal decomposition $\zeta = \iota \oplus \overline{\chi} = \iota \oplus \chi^*$ (cf. (vi) of §7.3), and a fixed isometric isomorphism $\chi^3 \to \Lambda^2\iota$ of Hermitian line bundles, which in turn is nothing else than a unit section Φ_0 of $(\Lambda^2\iota)\overline{\chi}^3$ (notation of (5.1)). This geometric structure in ζ can be characterized by selecting all orthonormal bases ψ, ϕ, ξ of each fibre ζ_x with $\psi, \phi \in \iota_x, \xi \in \overline{\chi}_x$ and $(\psi \wedge \phi)\xi\xi\xi = \Phi_0(x)$ (notation of (5.4)). These bases clearly form a G-structure in ζ, where $G \subset U(3)$ is the group of all matrices $(a, z) \in U(2) \times U(1) \subset U(3)$ such that $z^3 \det a = 1$. (Thus, G is a 3-fold covering group of U(2), with the projection homomorphism $(a, z) \mapsto a$.)

Remark 9.4.2.   For ι, ζ and their geometries as in Remark 9.4.1 we have a natural isomorphism $\mathcal{C}(\zeta) \to \mathcal{C}(\iota)$, which in terms of sections, i.e., compatible connections, is given by the restriction

$$\nabla \mapsto \nabla^{[\iota]} \tag{9.16}$$

obtained using the projection $\zeta = \iota + \overline{\chi} \to \iota$ as in (ix) of §6.6. Note that (9.16) is bijective since a compatible connection in ι determines one in $\lambda = \Lambda^2\iota$ and, consequently, one in $\overline{\chi} = \lambda^{-1/3}$ (see (x) of §7.3). The isomorphism in question can be made isometric for any given natural fibre metric in $\mathcal{C}(\iota)$ (or $\mathfrak{u}(\iota)$), and hence for any prescribed value of the Weinberg angle (§6.5, §6.8), if a suitable natural fibre metric is chosen in $\mathcal{C}(\zeta)$ (i.e., in $\mathfrak{g}(\zeta)$).

Remark 9.4.3.   Interacting-fermion bundles α other than (9.14) may also lead, via the symmetry-breaking substitution $\iota = 1 + \lambda$, to the decomposition

(9.15). For instance, the last summand in (9.14) could be replaced by either of

(a) $\lambda^{1/3}\bar{\iota}\sigma$, which gives rise to (9.15) whenever $\iota = 1 + \lambda$, i.e., $\bar{\iota} = 1 + \bar{\lambda} = 1 + \lambda^{-1}$; or,

(b) $\lambda^{-2/3}\iota\sigma_L + \lambda^{-2/3}\sigma_R + \lambda^{1/3}\sigma_R$. The resulting choice of α is preferred by physicists, as it enables one to extend to quarks not only the geometry of the electroweak model (as described above), but also its dynamics (§11.5).

References: Leite Lopes 1981; Ryder 1986.

10 Lagrangian in Classical Field Theory*

This chapter presents some basic facts of classical field theory in its Lagrangian formulation, including Lagrangian descriptions of energy and interactions.

The choice of material has been dictated by the applications in Chapter
11.

10.1 Lagrangian Densities

Most of the field equations encountered in particle physics are derived from
Lagrangian densities (briefly, Lagrangians), defined below. For exceptions,
see Remark 10.1.8.

A *Lagrangian* (*density*) in a fibre bundle η (which need not be a vector
bundle) over a given spacetime $(\mathcal{M}, g)$, is a first-order differential operator
$\mathcal{L}$, usually nonlinear, sending C^∞ sections ψ of η to C^∞ functions $\mathcal{L}(\psi)$:
$\mathcal{M} \to \mathbb{R}$. The *field equations* corresponding to $\mathcal{L}$ then consist in requiring
that a given C^∞ section ψ of η make the *U-action functional*

$$\psi \mapsto \int_U \mathcal{L}(\psi) \, dv_g \tag{10.1}$$

stationary for each open relatively compact set $U \subset \mathcal{M}$, with respect to C^∞
variations of ψ that are compactly supported in U, dv_g being the Riemannian measure of g. One easily sees that the field equations, also known as
the *Euler-Lagrange equations* for $\mathcal{L}$, can be written as the system

$$\frac{\partial \widetilde{\mathcal{L}}}{\partial \psi^a} - \partial_\mu \frac{\partial \widetilde{\mathcal{L}}}{\partial \psi^a_\mu} = 0 \tag{10.2}$$

of partial differential equations of order not exceeding two, imposed on ψ.
Here one uses a local trivialization of η along with local coordinates ψ^a
in its fibre and a local coordinate system x^μ in $\mathcal{M}$ with $\widetilde{\mathcal{L}} = \varphi \mathcal{L}$ and the
function φ given by $dv_g = \varphi \, dx^1 \dots dx^4$, regarding $[\widetilde{\mathcal{L}}(\psi)](x)$ as a function
of the $\psi^a = \psi^a(x)$, $\psi^a_\mu = \partial_\mu \psi^a(x)$ with $\partial_\mu = \partial/\partial x^\mu$, and x^μ.

The requirement that the given field equations be obtained as above
from a Lagrangian can be justified by analogy with classical mechanics, cf.
Remark 10.1.1 below. Although such a Lagrangian then is far from unique,

its "most natural" choice can often be made using plausible physical and formal arguments (Remark 10.1.2).

As an example, in the case where η is a real or complex vector bundle over $(\mathcal{M}, g)$, with a fixed, possibly indefinite, Riemannian or Hermitian fibre metric $\langle,\rangle$ and a connection $\overset{\circ}{\nabla}$ compatible with $\langle,\rangle$ (cf. §5.3), the natural Lagrangian

$$\mathcal{L}(\psi) = -\frac{1}{2}\left[\langle \overset{\circ}{\nabla}\psi, \overset{\circ}{\nabla}\psi \rangle + \frac{m^2 c^2}{\hbar^2} \langle \psi, \psi \rangle \right] \tag{10.3}$$

leads to the Klein-Gordon equation (1.3), where $\langle,\rangle$ also stands for the fibre metric in $\eta \otimes T^*\mathcal{M}$ induced by $\langle,\rangle$ and g. Furthermore, Dirac's equation (1.7) can be obtained from the Lagrangian

$$\mathcal{L}(\psi) = \frac{1}{2}\mathrm{Im}\,\omega(\psi, (\mathcal{D} + \frac{mc}{\hbar})\bar{\psi}) \tag{10.4}$$

in the given Dirac spinor bundle σ over $(\mathcal{M}, g)$, while the Weyl equation (1.6) may be derived from

$$\mathcal{L}(\psi) = \frac{1}{2}\mathrm{Im}\,\omega(\psi, \mathcal{D}\psi), \tag{10.5}$$

where ψ is a section of σ_L or σ_R for a fixed orientation of $\mathcal{M}$ (notation of §6.3). A natural Lagrangian for the Yang-Mills equation (5.7) in the affine bundle $\mathcal{C}(\delta)$ the sections of which are connections ∇ in the vector bundle δ that are compatible with some given geometry (G-structure) is

$$\mathcal{L}(\nabla) = -\frac{1}{4}\langle R^\nabla, R^\nabla \rangle, \tag{10.6}$$

provided that the geometry of δ also includes a natural fibre metric in $\mathfrak{g}(\delta)$ (cf. §6.8) needed, along with g, to make sense of $\langle,\rangle$ in (10.6). In the special case where $\delta = \lambda$ is the electromagnetism bundle (§4.10, §6.1), it is convenient to multiply (10.6) by a constant factor, obtaining the Lagrangian

$$\mathcal{L}(\nabla) = -\frac{1}{4\mu_0}\langle F, F \rangle, \tag{10.7}$$

with F given by (4.24) and μ_0 as in §4.1, i.e.,

$$\mathcal{L}(\nabla) = -\frac{\hbar^2}{4q^2 \mu_0}\langle R^\nabla, R^\nabla \rangle, \tag{10.8}$$

(see Example 10.2.3), which leads to the vacuum Maxwell equations (4.26).

We will also need a Lagrangian for matter particles of spin 1 and negative parity, described by real or complex valued 1-forms on $\mathcal{M}$, i.e., sections ψ of $T^*\mathcal{M}$ or its complexification $T^*\mathcal{M} \otimes \mathbb{C}$. According to §1.17, the field equations then should be

$$\left(\Box - \frac{m^2 c^2}{\hbar^2}\right)\psi = 0, \qquad \mathrm{div}\,\psi = 0, \tag{10.9}$$

$m > 0$ being the mass of the particle. However, one usually replaces (10.9) by *Proca's equation*

$$\mathrm{div}\,(d\psi) + \frac{m^2 c^2}{\hbar^2}\psi = 0, \tag{10.10}$$

having some desirable formal properties (see Remark 10.1.7), and equivalent to (10.9) in the case where $(\mathcal{M}, g)$ is the Minkowski spacetime M (or, more generally, whenever the Ricci tensor Ric of $(\mathcal{M}, g)$ vanishes identically, cf. Remark 10.1.6). A Lagrangian that gives rise to (10.10) can be defined by

$$\mathcal{L}(\psi) = -\frac{1}{4}g(d\psi, d\bar\psi) - \frac{1}{2}\frac{m^2 c^2}{\hbar^2}g(\psi, \bar\psi) \tag{10.11}$$

for 1-forms ψ on $\mathcal{M}$, with $\bar\psi = \psi$ if ψ is real. (See Remarks 10.1.4, 10.1.5 for notational conventions.)

Remark 10.1.1. The requirement that the given field equations in a fibre bundle η be obtained from a Lagrangian $\mathcal{L}$ is based on analogy with classical mechanics (cf. §4.7). In fact, in the case where η is a vector bundle, interpreting sections ψ of η as in (a) of §1.2 (with the appropriate assumptions on $\mathcal{M}$ and η), one may think of the elastic medium filling the 3-manifold ("space") $\mathbb{M}$ as being approximated by crystalline structures consisting of finitely many pointlike "atoms". Each such structure is expected to evolve in a manner determined by a Lagrangian $L = \sum_{\mathbf{x}} L_{\mathbf{x}}$ consisting of contributions $L_{\mathbf{x}}$ corresponding to the individual atoms $\mathbf{x}$, so that, for short time intervals $[a, b]$, the *action*

$$\int_a^b L\,dt = \sum_{\mathbf{x}}\int_a^b L_{\mathbf{x}}\,dt \tag{10.12}$$

is minimized among all trajectories of the system with prescribed endpoints. In the continuous-medium limit, it is clearly plausible to postulate that (10.12) approaches (10.1) for a suitable $U \subset \mathcal{M}$, which may then be assumed open and relatively compact due to the natural restrictions imposed on variations of ψ. The dependence of $L_{\mathbf{x}}$ on the state (position and velocity) of the finite dimensional system, as well as $\mathbf{x}$ and the time t, now indicates that $(\mathcal{L}(\psi))(x)$ should depend on ψ^a, $\partial\psi^a/\partial t$ and, separately, on x^μ (notations as in (10.2) with $x^0 = ct$ and coordinates x^j in $\mathbb{M}$, $j = 1, 2, 3$) while the possibility that $L_{\mathbf{x}}$ is also affected by the positions of the nearest neighbor atoms of $\mathbf{x}$ (as when a short-range interatomic force is involved) leads to a dependence of $(\mathcal{L}(\psi))(x)$ on the partial derivatives (up to some order) of the ψ^a relative to coordinates x^j in $\mathbb{M}$. Thus, $\mathcal{L}$ is usually a differential operator, often assumed for simplicity to be of the first order.

Remark 10.1.2. A Lagrangian $\mathcal{L}$ generating the prescribed field equations in a given fibre bundle η over $(\mathcal{M}, g)$ is far from unique, as it may be replaced by any operator of the form $\psi \mapsto \mathcal{L}(\psi) + \operatorname{div} \mathcal{A}(\psi)$ with a (possibly nonlinear) C^∞ bundle map $\mathcal{A} : \eta \to T\mathcal{M}$. In fact, this becomes clear when compactly supported variations are applied to (10.1). It is nevertheless possible, in most relevant cases, to select one Lagrangian in the huge class just described, so as to achieve maximum formal simplicity (the "least possible number of terms").

Additional freedom in choosing $\mathcal{L}$ comes from the possibility of multiplying it by nonzero real numbers. See also Remarks 10.1.3, 10.2.6.

Remark 10.1.3. In view of (4.11) we expect all Lagrangians $\mathcal{L}$ to represent the physical units of (spatial) energy density, cf. §10.2. In (10.3) – (10.6) and (10.11) this can be achieved by using suitable multiplicative factors and/or assuming that the components of ψ have a specific physical dimension. In most cases, however, one leaves the Lagrangian formulae (10.3) – (10.6), (10.11) unchanged, ignoring the problem of units, as it does not lead to serious confusion.

Remark 10.1.4. For k-forms α, β on $(\mathcal{M}, g)$ valued in a vector bundle with a fixed Riemannian or Hermitian fibre metric $\langle , \rangle$ we set

$$\langle \alpha, \beta \rangle = g^{\mu_1 \nu_1} \cdots g^{\mu_k \nu_k} \langle \alpha_{\mu_1 \ldots \mu_k}, \beta_{\nu_1 \ldots \nu_k} \rangle,$$

where $\alpha_{\mu_1 \ldots \mu_k} = \alpha(\partial_{\mu_1}, \ldots, \partial_{\mu_k})$ and $\partial_\mu = \partial / \partial x^\mu$ in any local coordinates x^μ for $\mathcal{M}$, i.e.,

$$\langle \alpha, \beta \rangle = k! \sum_{\mu_1 < \ldots < \mu_k} \varepsilon_{\mu_1} \cdots \varepsilon_{\mu_k} \langle \alpha(e_{\mu_1}, \ldots, e_{\mu_k}), \beta(e_{\mu_1}, \ldots, e_{\mu_k}) \rangle$$

whenever a basis e_μ of $T_x\mathcal{M}$, $x \in \mathcal{M}$, is orthonormal in the sense that $\langle e_\mu, e_\nu \rangle = \varepsilon_\mu \delta_{\mu\nu}$, $\varepsilon_\mu = \pm 1$. This applies in particular to forms valued in complex, real or imaginary numbers, with $\langle z, u \rangle = \operatorname{Re} z\overline{u}$ for such numbers z, u, in which case we also write

$$\langle \alpha, \beta \rangle = g(\alpha, \overline{\beta}). \tag{10.13}$$

Remark 10.1.5. Our notations in (10.10) are such that

$$\operatorname{div}(d\psi) = [\nabla^\mu (\partial_\nu \psi_\mu - \partial_\mu \psi_\nu)] dx^\nu$$

for a 1-form $\psi = \psi_\mu dx^\mu$ on $(\mathcal{M}, g)$, x^μ being any local coordinates in $\mathcal{M}$.

Remark 10.1.6. Proca's equation (10.10) with positive constants $m, c, \hbar$ for a 1-form ψ on a pseudo-Riemannian manifold $(\mathcal{M}, g)$ (of any dimension) is equivalent to the system

$$\left(\Box - \frac{m^2 c^2}{\hbar^2}\right)\psi = \mathrm{Ric}\,(\psi), \quad \mathrm{div}\,\psi = 0,$$

as one easily concludes using the *Ricci-Weitzenböck formula* $\mathrm{Ric}(\psi) = \mathrm{div}\,(d\psi) + \Box\psi - d(\mathrm{div}\,\psi)$ and the fact that $\mathrm{div} \circ \mathrm{div} = 0$ (notations of Remark 10.1.5 and §1.16 with $\mathrm{div}\,\psi = \mathrm{Trace}\,(\nabla\psi)$).

Remark 10.1.7. The use of Proca's equation (10.10) instead of the system (10.9) is justified by the fact that, in the vanishing-mass limit $m \to 0$, massive particles described by (10.10) naturally "approach" massless particles of the same spin and parity, namely photons. In fact, formal symmetry breaking applied to the electromagnetism bundle λ (§6.1) allows us to replace compatible connections ∇ in λ by their connection forms Γ with (4.22). The free-photon field equations, i.e., the vacuum Maxwell equations (4.26) then can be rewritten as (10.10) with $m = 0$ and $\psi = \Gamma$ (cf. (4.23)). Also, up to a factor (see Remark 10.2.4), the photon Lagrangian (10.7) corresponds to (10.11) with $m = 0$.

Remark 10.1.8. The field equations (5.19) in a harmonic line bundle λ are *not* of the form (10.2) for any Lagrangian $\mathcal{L}$. Therefore, a Lagrangian for the free photon, described as in §5.11, has to be selected so as to lead to some *weaker* field equations, while the full strength of (5.19) then is regained only by imposing a *subsidiary condition* on ∇. Besides the standard Lagrangian (10.6) (or (10.8)), for which (10.2) is the Maxwell equation $\mathrm{div}\,R^\nabla = 0$, one also uses here the Lagrangian

$$\mathcal{L}(\nabla) = -\frac{1}{4}\langle R^\nabla, R^\nabla \rangle - \frac{1}{2}(\mathrm{div}\,\nabla)^2$$

which, by (5.17), gives rise to the field equation

$$(\Box - \mathrm{Ric})\nabla = 0,$$

i.e., the Klein-Gordon equation with the Ricci term. In both cases, one may choose the subsidiary condition to be $\mathrm{div}\,\nabla = 0$.

References: Binz, Śniatycki and Fischer 1988; Ramond 1981; Bogolyubov and Shirkov 1983; DeWitt 1965.

10.2 Energy Density

Given a Lagrangian $\mathcal{L}$ in a fibre bundle η over $(\mathcal{M}, g)$ as in §10.1, let us assume, in addition, that there are fixed identifications

$$\mathcal{M} = \mathbb{R} \times \mathbb{M}, \quad dv_g = dt \cdot dv_{\mathbb{M}}, \quad \eta = \mathrm{pr}^* \zeta \tag{10.14}$$

(cf. §1.2), where ζ is a bundle over the 3-manifold $\mathbb{M}$ endowed with a smooth positive measure $dv_{\mathbb{M}}$ and $\mathrm{pr} : \mathbb{R} \times \mathbb{M} \to \mathbb{M}$ is the projection, while t (interpreted as time) is the standard coordinate on $\mathbb{R}$ and so dt is the Lebesgue measure. Using a coordinate system in $\mathcal{M}$ consisting of t and some local coordinates in $\mathbb{M}$, along with local coordinates ψ^a in the fibre of ζ and a local trivialization of η obtained via (10.14) from one for ζ, we can define the real-valued C^∞ function

$$\mathcal{H}(\psi) = \dot{\psi}^a \frac{\partial \mathcal{L}}{\partial \dot{\psi}^a} - \mathcal{L}(\psi) \tag{10.15}$$

for any C^∞ section ψ of η, where $\dot{\psi}^a = \partial \psi^a / \partial t$. Obviously, $\mathcal{H}(\psi)$ depends only on ψ, $\mathcal{L}$ and the identifications (10.14). Thus, despite the *local* character of formula (10.15), involving a trivialization and coordinates, $\mathcal{H}(\psi)$ is well-defined everywhere in $\mathcal{M}$. In view of (4.11) it is natural to interpret $\mathcal{H}(\psi)$ as the *energy density* function of the given system (particle) in state ψ. The assignment $\psi \mapsto \mathcal{H}(\psi)$ is a (usually nonlinear) first order differential operator, depending on the choice of the "observer" (10.14).

In the following examples we assume, in addition, that relations (10.14) are *compatible* with g, i.e.,

$$g = (-dt^2) \times g_{\mathbb{M}}, \quad dv_{\mathbb{M}} = dv_{g_{\mathbb{M}}}. \tag{10.16}$$

In other words, we have a Riemannian-product decomposition $(\mathcal{M}, g) = (\mathbb{R}, -dt^2) \times (\mathbb{M}, g_{\mathbb{M}})$ with a Riemannian metric $g_{\mathbb{M}}$ on $\mathbb{M}$ whose Riemannian measure is $dv_{\mathbb{M}}$. It is also convenient to introduce the Riemannian metric $g_+ = dt^2 \times g_{\mathbb{M}}$ on $\mathcal{M}$ and denote $\langle , \rangle_+$ the inner product of forms on $\mathcal{M}$ valued in vector bundles, defined as in Remark 10.1.4 with g_+ instead of g.

Example 10.2.1. Let, moreover, η be a vector bundle over $\mathcal{M}$ endowed with a fibre metric $\langle , \rangle$ and a compatible connection $\overset{\circ}{\nabla}$, both obtained as pull-backs under pr of analogous objects in the vector bundle ζ over $\mathbb{M}$. The Lagrangian (10.3), leading to the Klein-Gordon equation (1.3) with the mass m, then gives rise to the energy density

$$\mathcal{H}(\psi) = \frac{1}{2} \left[\langle \overset{\circ}{\nabla} \psi, \overset{\circ}{\nabla} \psi \rangle_+ + \frac{m^2 c^2}{\hbar^2} \langle \psi, \psi \rangle \right]. \tag{10.17}$$

Example 10.2.2. In the case of the Yang-Mills Lagrangian (10.6) in the affine bundle $\eta = \mathcal{C}(\delta)$ over $\mathcal{M}$, the energy density given by (10.15) fails to have three natural, highly desirable properties, listed in Remark 10.2.8 below (even if the vector bundle δ and its G-structure are pull-backs of similar objects over IM). Therefore, for Yang-Mills fields one replaces (10.15) by the *modified* energy density formula

$$\mathcal{H}(\nabla) = \frac{1}{4}\langle R^\nabla, R^\nabla \rangle_+ , \tag{10.18}$$

satisfying all the required conditions, and obtained as in (10.15) using natural "velocity components" different from the $\dot{\psi}^a$. See Remarks 10.2.8, 10.2.9 for details.

Example 10.2.3. If δ in Example 10.2.2 happens to be the electromagnetism bundle λ (§6.1), the Lagrangian formula (10.8) contains, in contrast with (10.6), an additional factor of $\hbar^2 q^{-2}/\mu_0$. This is justified not only by $\mathcal{L}(\nabla)$ then having the correct physical dimension of energy density but, first of all, by the fact that the resulting expression

$$\mathcal{H}(\nabla) = \frac{\hbar^2}{4q^2\mu_0}\langle R^\nabla, R^\nabla \rangle_+ = \frac{1}{4\mu_0}\langle F, F \rangle_+$$

coincides, for the Minkowski spacetime with a fixed observer compatible with (10.14), (10.16), with the electromagnetic energy-density formula

$$\mathcal{H} = \frac{\epsilon_0}{2}|\mathbf{E}|^2 + \frac{1}{2\mu_0}|\mathbf{B}|^2 \tag{10.19}$$

in classical electrodynamics (cf. (4.5)), which one derives without any reference to Yang-Mills fields.

Remark 10.2.4. In cases where photons are discussed simultaneously with other, similar particles, the additional factor in (10.8) is sometimes dropped to make the approach more uniform. See, e.g., Remarks 10.1.7, 11.5.2.

Remark 10.2.5. The total relativistic energy of any object is nonnegative, since p in (4.7) points toward future. It is therefore natural to expect the energy density $\mathcal{H}$ with (10.15) to be *positive semidefinite* in the sense that $\mathcal{H}(\psi) \geq 0$ for all C^∞ sections ψ of η. This is for instance the case in Examples 10.2.1, 10.2.2 above (provided that the fibre metric $\langle , \rangle$ in η or $\mathfrak{g}(\delta)$ is positive definite), and in Example 10.2.3.

Remark 10.2.6. Our freedom of modifying Lagrangians that lead to given field equations, using constant factors (Remark 10.1.2), is restricted by the energy-density interpretation of (10.15). Thus, if the positivity requirement of Remark 10.2.5 is to remain satisfied, the factor must be positive. In cases such as Example 10.2.3, where an energy-density formula can be derived independently of the Lagrangian, the freedom disappears completely, i.e., the only possible factor is 1.

Remark 10.2.7. The energy density corresponding under the assumptions (10.14), (10.16) to the Lagrangian $\mathcal{L}$, given by (10.4), and leading to Dirac's equation, is not positive (or negative) semidefinite. In view of Remark 10.2.5, this is a serious flaw of the Dirac equation as a description of an elementary particle, illustrating how inadequate a classical approach may be. Positivity of the relativistic energy can be restored in this case by applying the fermion version of field quantization, or, still on the classical level, by tensoring the Dirac spinors with suitable "anticommuting numbers".

Remark 10.2.8. The following conditions are satisfied by (10.18), but not by the original energy density (10.15) for Yang-Mills fields:
(a) gauge invariance (see §5.9),
(b) nonnegativity, whenever the fibre metric in $\mathfrak{g}(\delta)$ is positive definite (cf. Remark 10.2.5),
(c) relation (10.19) in the case of electromagnetism over the Minkowski spacetime.

Remark 10.2.9. Formula (10.18) is a natural gauge-invariant modification (cf. §5.9) of (10.15) for Yang-Mills fields. In fact, $\mathcal{L}$ in (10.6) is, locally, a function of $x \in \mathcal{M}$ and the components of the curvature R^{∇}, the latter thus forming a convenient quasi-linear substitute for the partial derivatives of the connection components. This in turn suggests replacing $\dot{\psi}^a$ in (10.15) by those curvature components involving the time direction, i.e., the local components of the section $R^{\nabla}(\partial/\partial t, \cdot)$ of $\mathfrak{g}(\delta) \otimes T^*\mathcal{M}$, which immediately leads to (10.18).

References: Binz, Śniatycki and Fischer 1988; Huang 1982.

10.3 Interaction Lagrangians

Let the fibre bundles $\eta_1, \ldots, \eta_n$ over the given spacetime $(\mathcal{M}, g)$ represent n distinct particle species, each of which obeys some field equations derived as in §10.1 from a specific Lagrangian $\mathcal{L}_j$ in η_j, $j = 1, \ldots, n$. Allowing these particles, in addition, to interact with one another, it is natural to ask how a collection $(\psi_1, \ldots, \psi_n)$ of states of the individual particles, i.e., sections ψ_j of the η_j, is going to evolve under the influence of the evolution of each ψ_j (governed by its field equations) *and* their mutual interactions. Instead of directly describing the field equations to be imposed on $(\psi_1, \ldots, \psi_n)$, it is convenient to obtain them from a Lagrangian $\mathcal{L}$ in the bundle whose sections have the form $(\psi_1, \ldots, \psi_n)$ (see Remark 10.3.4 below). A general answer then can be stated as the *addition rule* for Lagrangians, which says that

$$\mathcal{L}(\psi_1, \ldots, \psi_n) = \sum_{j=1}^{n} \mathcal{L}_j(\psi_j) + \mathcal{L}_{\text{int}}(\psi_1, \ldots, \psi_n) \tag{10.20}$$

with a suitable *interaction term* $\mathcal{L}_{\text{int}}$, often given as a sum

$$\mathcal{L}_{\text{int}}(\psi_1, \ldots, \psi_n) = \sum_{j \neq k} \mathcal{L}_{\text{int}}^{j,k}(\psi_j, \psi_k) \tag{10.21}$$

of two-particle contributions. (See Remarks 10.3.5, 10.3.6.) The subsequent discussion is aimed at determining $\mathcal{L}_{\text{int}}$ in various specific situations.

Typically, among the $\psi_1, \ldots, \psi_n$ there will be pairs $(\psi_j, \psi_k) = (\phi, \nabla)$, where ϕ is a section of a natural vector bundle $\eta = \eta_j$ with the canonical connection $\overset{\circ}{\nabla}$ over $(\mathcal{M}, g)$ (see §1,.18), corresponding to a matter-particle species subject to an interaction represented by an interaction bundle δ (§5.4) with $\eta_k = \mathcal{C}(\delta)$, while $\nabla = \psi_k$ is a compatible connection in δ, interpreted as a state of an interaction-carrier particle. In this case we already have an idea how to account for the interaction in all expressions, including Lagrangians. Namely, we use the minimum-coupling substitution (5.11), which consists in replacing $\overset{\circ}{\nabla}$ and the section ϕ of η, wherever they occur in the original expression, by $\nabla \otimes \overset{\circ}{\nabla}$ and, respectively, a section ψ of $\delta \otimes \eta$ (or, of a more general interacting-particle bundle (5.12)). This is consistent with the addition rule (10.20), as one easily verifies using identifications based on selecting suitable local trivializing sections for δ, along with the fact that the relevant matter-particle Lagrangians $\mathcal{L}(\phi)$ are all polynomial in $\overset{\circ}{\nabla}\phi$, cf. (10.3) – (10.5), (10.11) and (1.4). (Also, note that for 1-forms ϕ on $(\mathcal{M}, g)$, $d\phi$ is a multiple of the skew-symmetric part of $\overset{\circ}{\nabla}\phi$, where $\overset{\circ}{\nabla}$ stands for the Levi-Civita connection of g.)

Therefore, for Yang-Mills type interactions, *the addition rule* (10.20) *along with the minimum-coupling substitution* (5.11) *provide a complete description of the interaction Lagrangian.*

Example 10.3.1. Given a vector bundle η with $\overset{\circ}{\nabla}$ as above and a free-particle Lagrangian $\mathcal{L}_{\text{free}}$ in η, we can always write

$$(\mathcal{L}_{\text{free}}(\phi))(x) = \mathcal{L}_0(\phi(x), (\overset{\circ}{\nabla}\phi)(x)) \tag{10.22}$$

for sections ϕ of η and a function $\mathcal{L}_0 : \eta \oplus (T^*\mathcal{M} \otimes \eta) \to \mathbb{R}$. Combining $\mathcal{L}_{\text{free}}$ with the Yang-Mills Lagrangian (10.6), we thus obtain the (*total*) *interaction Lagrangian*

$$\mathcal{L}(\psi, \nabla) = \mathcal{L}_0(\psi, (\nabla \otimes \overset{\circ}{\nabla})\psi) - \frac{1}{4}\langle R^\nabla, R^\nabla \rangle \tag{10.23}$$

for sections ψ of $\delta \otimes \eta$ (or (5.12)) and compatible connections ∇ in δ, assuming, of course, that $\mathcal{L}_0$ is "general" enough to make sense also for $\delta \otimes \eta$ instead of η.

The Lagrangian (10.23) leads to the coupled field equations required in §5.4. In fact, direct computation shows that the field equations (10.2) corresponding to (10.23) in the affine bundle $(\delta \otimes \eta) \oplus C(\delta)$ are (5.9) and (5.10), where $\mathcal{F}$ is such that (5.8) are the field equations derived from $\mathcal{L}_{\text{free}}$, div is defined using the same sign convention as in §4.2, C_0 is any fixed nonzero constant, while $J(\psi)$ is a section of $TM \otimes \operatorname{End} \delta$, characterized as follows. Fix $x \in M$ and let F be the differential at $((\nabla \otimes \overset{\circ}{\nabla})\psi)(x) \in T_x^* \delta_x \eta_x$ (notation of §5.1 with $T = TM$) of the function $T_x^* \delta_x \eta_x \ni a \mapsto \mathcal{L}_0(\psi(x), a) \in \mathbb{R}$, so that $F \in T_x \delta_x^* \eta_x^*$. Then $(J(\psi))(x)$ equals C_0^{-1} times the image of $F \otimes \psi(x)$ under the vector bundle morphism $(T\delta^* \eta^*)(\delta\eta) \to T\delta^* \delta = TM \otimes \operatorname{End} \delta$ induced in an obvious manner by the pairing morphism $\eta^* \eta \to 1$.

Example 10.3.2. As a special case of Example 10.3.1, let us consider a Dirac spinor bundle $\eta = \sigma$ over (M, g) with the Lagrangian $\mathcal{L}_{\text{free}}$ given by (10.4), leading to Dirac's equation (1.7), and an electromagnetism bundle λ corresponding to some amount $q \neq 0$ of the electric charge, which does not have to be the charge of the electron, with the Lagrangian (10.8) (rather than (10.6)), that gives rise to the vacuum Maxwell equations (4.26). The total interaction Lagrangian $\mathcal{L}(\psi, \nabla) = \frac{1}{2}\operatorname{Im}\omega(\psi, \mathcal{D}^\nabla \bar{\psi}) - \frac{\hbar^2}{4q^2 \mu_0}\langle R^\nabla, R^\nabla \rangle$ (with $\mathcal{D}^\nabla$ as in Remark 5.4.2) then leads to the coupled field equations

$$\left(\mathcal{D}^\nabla + \frac{mc}{\hbar}\right)\psi = 0, \quad \operatorname{div} R^\nabla = \frac{1}{2}i\mu_0 q^2 \hbar^{-2} \operatorname{Im} w_\psi \tag{10.24}$$

where w_ψ is the complex vector field on M (i.e., section of $TM \otimes \mathbb{C}$) characterized by $w_\psi(p) = \omega(\psi, \gamma(p)\bar{\psi})$ whenever $p \in T_x^* M$ and $x \in M$ (see §1.3). Equations (10.24) describe the simultaneous "classical" evolution of a particle of spin $\frac{1}{2}$, mass m and charge q, and a photon, subject to a mutual electromagnetic interaction and represented, respectively, by sections ψ of $\lambda \otimes \eta$ and Hermitian connections ∇ in λ.

Example 10.3.3. Another application of Example 10.3.1 can easily be obtained for the product line bundle $\eta = M \times \mathbb{C}$ and the Lagrangian (10.3) in η (where $\overset{\circ}{\nabla} = d$ is the standard flat connection), which gives rise to the Klein-Gordon equation (1.3). With $\rho = \lambda$ and q as in Example 10.3.2, we obtain the coupled field equations

$$\left(\Box - \frac{m^2 c^2}{\hbar^2}\right)\psi = 0, \quad \operatorname{div} R^\nabla = -i\mu_0 q^2 \hbar^{-2} \operatorname{Im} v_\psi \tag{10.25}$$

where ψ is a section of $\tilde{\lambda} = \lambda \otimes \eta$ and ∇ is a Hermitian connection in λ representing, respectively, states of a given particle with mass m, spin 0, positive parity and charge q, subject to the electromagnetic interaction, and those of the photon. Here v_ψ is the complex vector field on M with $g(v_\psi, u) = \langle \nabla_u \psi, \psi \rangle$ for $u \in T_x M$ and $x \in M$, $\langle, \rangle$ being the Hermitian fibre metric in λ, while the d'Alembertian $\square$ in λ, appearing in (10.25), is formed using ∇ (cf. §1.16).

Remark 10.3.4. Given sections ψ_j of fibre bundles η_j over $\mathcal{M}$, $j = 1, \ldots, n$, the n-tuple $(\psi_1, \ldots, \psi_n)$ can be regarded as a section of the "direct sum" bundle $\eta_1 \oplus \ldots \oplus \eta_n$ whose fibre over each $x \in \mathcal{M}$ is $(\eta_1)_x \times \ldots \times (\eta_n)_x$.

Remark 10.3.5. The addittition rule (10.20) for Lagrangians can be justified by the linear relation (10.15) between any Lagrangian $\mathcal{L}$ and the corresponding energy density $\mathcal{H}$ (which makes sense under the assumption (10.14)), and the fact that a similar additivity property for the energy is intuitively obvious. For instance, the total (mechanical) energy of a system of celestial bodies is obtained by summing the kinetic energies of the constituents and the potential (interaction) energies, due to gravity, of all two-body subsystems.

Remark 10.3.6. The addition rule is further made plausible by the fact that, if the interaction is absent and so we may set $\mathcal{L}_{\text{int}} = 0$ in (10.20), the resulting Lagrangian

$$\mathcal{L}(\psi_1, \ldots, \psi_n) = \sum_j \mathcal{L}_j(\psi_j) \tag{10.26}$$

in $\eta_1 \oplus \ldots \oplus \eta_n$ leads to the original, uncoupled field equations, i.e., all particle species in question evolve as before, ignoring one another's presence.

Remark 10.3.7. An interaction is called *pointlike* if it takes place directly between matter particles, without being aided by interaction carriers (represented in the Yang-Mills formalism by connections). Yang-Mills interactions themselves may appear pointlike if one uses symmetry breaking (§5.8) to identify some or all of their carriers with matter particles. The extent to which such a procedure is physically relevant depends on whether the breaking of symmetry is formal, or actually takes place in nature.

The latter case is illustrated by the electroweak model (§6.5), where the weak interaction carriers $W^\pm$, Z^0 are obtained via spontaneous symmetry breaking that leads to their description as matter fields (see statement (ii) following (6.30)). As a consequence, the weak interaction can reasonably be regarded as pointlike (cf. the first paragraph of §6.5). On the other hand, pointlike interactions that are *not* obtained by applying symmetry breaking to Yang-Mills fields, are also useful as a theoretical concept and may even correspond to physical reality (see (c) of §11.5).

Remark 10.3.8. It is natural to expect that the interaction term which has to appear in the total Lagrangian for several particle species in order to account for a given pointlike interaction (Remark 10.3.7), is a homogeneous, fibrewise polynomial function of (the 1-jet of) each section ψ involved (as well as $\bar{\psi}$, if the vector bundle where ψ lives is complex). In fact, this is immediate in the case where symmetry breaking in its most drastic form (trivialization) is imposed on the minimum-coupling substitution, replacing Yang-Mills connections interacting with matter by their connection forms, in view of the polynomial dependence of the standard matter-field Lagrangians (10.3) – (10.5), (10.11) on $\overset{\circ}{\nabla}\psi$. The same is true for the (cubic and quartic) *self-interaction terms* in the Lagrangian (10.6) of non-Abelian Yang-Mills fields, cf. §5.4.

References: Bleecker 1981; Leite Lopes 1981.

11 Symmetry Breaking and Positive Masses*

This chapter describes how particle masses can be *defined* in terms of a given Lagrangian density along with a fixed energy-minimizing *vacuum state*. We then proceed to discuss the dynamics of the electroweak model, using the *Higgs mechanism*. The latter consists in a refined version of the definition of mass that leads to the required massive particles e, W^+, W^-, but prevents the appearance of any unwanted, massless *Goldstone bosons*.

11.1 Vacuum States

Let $\mathcal{L}$ be a Lagrangian (§10.1) in a fibre bundle η over a spacetime $(\mathcal{M}, g)$, admitting a fixed decomposition (10.14). A C^∞ section ψ_0 of η is said to be a *vacuum state* for $\mathcal{L}$ relative to the decomposition (10.14) if it provides a pointwise minimum for the corresponding energy density $\mathcal{H}$ with (10.15), i.e., if $\mathcal{H}(\psi_0) \leq \mathcal{H}(\psi)$, everywhere on $\mathcal{M}$, for all C^∞ sections ψ of η.

In the following examples, we assume (10.14) and (10.16), except in Example 11.1.4.

Example 11.1.1. Let $\mathcal{L}$ be the Lagrangian (10.3), leading to the Klein-Gordon equation (1.3) with mass m, in a real or complex vector bundle η with a *positive definite* Riemannian or Hermitian fibre metric $\langle , \rangle$ and a compatible connection $\overset{\circ}{\nabla}$ satisfying the conditions listed in Example 10.2.1. The zero section of η then is the only vacuum state for $\mathcal{L}$, as $\langle , \rangle_+$ and $\langle , \rangle$ in (10.17) are both positive definite.

Example 11.1.2. For the Yang-Mills Lagrangian (10.6) in the affine bundle $\eta = \mathcal{C}(\delta)$, where δ is a vector bundle with a G-structure and a *positive definite* natural fibre metric in $\mathfrak{g}(\delta)$, the only vacuum states are, by (10.18), those compatible connections ∇ in δ which are flat. (Their existence imposes obvious topological restrictions on δ and the G-structure.)

Example 11.1.3. For $\eta, \overset{\circ}{\nabla}, \delta$ and $\mathcal{L}_{\text{free}}$ as in Example 10.3.1, the corresponding energy-density operator $\mathcal{H}_{\text{free}}$ obviously satisfies the analogue of (10.22) with some function $\mathcal{H}_0$, assumed here to be as "general" as $\mathcal{L}_0$. The energy density $\mathcal{H}$ for the interaction Lagrangian (10.23) in $(\delta \otimes \eta) \oplus \mathcal{C}(\delta)$ then is given by

$$\mathcal{H}(\psi, \nabla) = \mathcal{H}_0(\psi, (\nabla \otimes \overset{\circ}{\nabla})\psi) + \frac{1}{4}\langle R^\nabla, R^\nabla \rangle_+ \tag{11.1}$$

(cf. (10.15), (10.18)). As a consequence, in the case where the fibre metric in $\mathfrak{g}(\delta)$ is positive definite, while $\mathcal{H}_0 \geq 0$ and $\mathcal{H}_0(\psi, \psi') = 0$ only if $\psi = 0$ and $\psi' = 0$, the only vacuum states in $(\delta \otimes \eta) \oplus \mathcal{C}(\delta)$ are the pairs $(0, \nabla)$ formed by the zero section of $\delta \otimes \eta$ and any flat connection ∇ in δ compatible with its geometry (provided that such connections exist).

Example 11.1.4. The descriptions of vacuum states in the above examples contain no reference to the decomposition (10.14). Thus, even without assuming (10.14) (or any positivity condition), we may still formally define a *vacuum state* for a Lagrangian $\mathcal{L}$ in a fibre bundle over a spacetime $(\mathcal{M}, g)$ to be

(a) The zero section, if $\mathcal{L}$ is one of the quadratic Lagrangians (10.3) – (10.5), (10.11) in the appropriate vector bundle;

(b) The zero section, if $\mathcal{L}$ is the interaction-free, direct-sum Lagrangian (10.26) in the direct sum $\eta_1 \oplus \ldots \oplus \eta_n$ of several copies of the vector bundles mentioned in (a), with some masses m_j, equal to zero for those j with $\mathcal{L}_j$ given by (10.5). Note that zero is the only section minimizing the energy density $\mathcal{H}$ of (10.26) with $\mathcal{H}(\psi_1 \ldots, \psi_n) = \sum_j \mathcal{H}_j(\psi_j)$, in the case where each $\mathcal{L}_j$ is as in Example 11.1.1 and (10.14), (10.16) hold for $\eta = \eta_j$, $j = 1, \ldots n$;

(c) Any flat compatible connection in δ, for $\mathcal{L}$ given by (10.6) in the affine bundle $\mathcal{C}(\delta)$, δ being a fixed vector bundle with a G-structure over $\mathcal{M}$;

(d) A section of $(\delta \otimes \eta) \oplus \mathcal{C}(\delta)$ having the form $(0, \nabla)$ with any flat compatible connection ∇ in δ, if η, δ and the interaction Lagrangian $\mathcal{L}$ are as in Example 10.3.1.

References: Bleecker 1981; Felsager 1983; Ryder 1985.

11.2 Lagrangians and Potentials

By a *connection* in a fibre bundle η over the given spacetime $(\mathcal{M}, g)$ we mean a subbundle χ of the tangent bundle $T\eta$ of the total space which is *horizontal*, i.e., $T\eta = \chi \oplus \upsilon$, υ being the *vertical* subbundle of $T\eta$ with $\upsilon_\phi = T_\phi \eta_x$ for $x \in \mathcal{M}$, $\phi \in \eta_x$. Note that, in the case where η is a vector bundle, this class of connections contains more than just the *linear* ones discussed in §5.3, as the corresponding parallel transports between fibres of η may be nonlinear.

A fixed connection χ in η associates with any Lagrangian $\mathcal{L}$ in η (§10.1) its *potential* (or *zero-order term*), which is a function $\mathcal{V} : \eta \to \mathbb{R}$, that is, a zero-order operator sending sections of η onto functions $\mathcal{M} \to \mathbb{R}$, defined as follows. Given $x \in \mathcal{M}$ and $\phi \in \eta_x$, we set $\mathcal{V}(\phi) = -[\mathcal{L}(\psi)](x)$, ψ being any C^∞ section of η with $\psi(x) = \phi$, defined near x and *parallel* at x in the sense that $d\psi(T_x\mathcal{M}) = \chi_\phi$. Note that $\mathcal{V}(\phi)$ does not depend on the choice of ψ, since $\mathcal{L}$ is a first-order differential operator.

In the case where χ corresponds to a linear connection ∇ in a vector bundle η, a section ψ of η is parallel at $x \in \mathcal{M}$ if and only if $(\nabla\psi)(x) = 0$. Therefore, the potential $\mathcal{V}$ of any Lagrangian $\mathcal{L}$ in η then is obtained by formally setting $\nabla\psi = 0$ in the formula for $-\mathcal{L}$. In particular, each of the Lagrangians (10.3)-(10.5), (10.11) in the appropriate vector bundle with the specific distinguished connection ($\overset{\circ}{\nabla}$, the spinor connection, or the Levi-Civita connection), gives rise to a potential $\mathcal{V}$ whose restriction to each fibre is *homogeneous quadratic*. More precisely, $\mathcal{V}$ in these examples may be regarded as a *mass term*, since it vanishes for the Lagrangian (10.5) (describing the massless neutrinos) and, in the remaining three cases,

$$\mathcal{V}(\psi) = \frac{1}{2}\left(\frac{mc}{\hbar}\right)^k \langle\psi,\psi\rangle, \tag{11.2}$$

where m is the mass of the corresponding particle and $\langle,\rangle$ stands for a natural fibre metric (cf. (6.7), (10.13)), while $k \in \{1,2\}$ is the order of the resulting field equation (or, equivalently, the degree of $\mathcal{L}$ as a polynomial in the first-order partial derivatives of ψ).

In the following remarks and examples, whenever energy density (or, potential) is mentioned, we automatically assume that the spacetime $(\mathcal{M}, g)$ and the fibre bundle η over $\mathcal{M}$ have a fixed decomposition (10.14) (or, respectively, that η carries a distinguished connection χ).

Remark 11.2.1. As in the case of finite dimensional conservative systems, it is often useful to represent a given Lagrangian $\mathcal{L}$ in η in the form $\mathcal{L} = \mathcal{K} - \mathcal{V}$, where $\mathcal{V}$ is the potential of $\mathcal{L}$, while $\mathcal{K} = \mathcal{L} + \mathcal{V}$ may be called its *kinetic term*. However, the analogy is mainly formal, since in the continuous-medium limit, the short-range interaction part of the potential energy in classical mechanics contributes to $\mathcal{K}$ rather than $\mathcal{V}$ (see Remark 10.1.1).

Remark 11.2.2. The assignments sending each Lagrangian $\mathcal{L}$ in η onto its energy density $\mathcal{H} = \mathcal{H}_\mathcal{L}$ with (10.15), and potential $\mathcal{V} = \mathcal{V}_\mathcal{L}$, are linear. Thus, regarding $\mathcal{K}$, $\mathcal{V}$ with $\mathcal{L} = \mathcal{K} - \mathcal{V}$, described in Remark 11.2.1, as Lagrangians in η, we have $\mathcal{H}_\mathcal{V} = -\mathcal{V}$, $\mathcal{H} = \mathcal{H}_\mathcal{L} = \mathcal{H}_\mathcal{K} + \mathcal{V}$, while $\mathcal{V}_\mathcal{K} = 0$ and $\mathcal{V}$ is its own potential.

Remark 11.2.3. By Remark 11.2.2, a Lagrangian $\mathcal{L}_1 = \mathcal{K} - \mathcal{V}_1$ in η (as in Remark 11.2.1) can always be replaced by the Lagrangian $\mathcal{L}_2 = \mathcal{K} - \mathcal{V}_2 = \mathcal{L}_1 + \mathcal{V}_1 - \mathcal{V}_2$ with the same kinetic term $\mathcal{K}$ and any prescribed potential $\mathcal{V}_2 : \eta \to \mathbb{R}$. The energy densities $\mathcal{H}_j$ of the $\mathcal{L}_j$ then are related by $\mathcal{H}_2 = \mathcal{H}_1 + \mathcal{V}_2 - \mathcal{V}_1$.

Example 11.2.4. Let $\overset{\circ}{\nabla}$ be a fixed (linear) connection compatible with a positive definite fibre metric $\langle,\rangle$ in a vector bundle η over $\mathcal{M}$, and let $f : [0,\infty) \to \mathbb{R}$ be a C^∞ function. As in Remark 11.2.3, we may replace the potential (11.2) (with $k = 2$) of the corresponding Klein-Gordon Lagrangian (10.3), by the new potential

$$\mathcal{V}(\phi) = f(\langle\phi,\phi\rangle) \tag{11.3},$$

thus obtaining the new Lagrangian

$$\mathcal{L}(\phi) = -\frac{1}{2}\langle\overset{\circ}{\nabla}\phi, \overset{\circ}{\nabla}\phi\rangle - f(\langle\phi,\phi\rangle) \tag{11.4}$$

with the potential (11.3), and the energy density, given by the formula

$$\mathcal{H}(\phi) = \frac{1}{2}\langle\overset{\circ}{\nabla}\phi, \overset{\circ}{\nabla}\phi\rangle_+ + f(\langle\phi,\phi\rangle) \tag{11.5}$$

obtained from (10.17) and Remark 11.2.3, whenever (10.14), (10.16) are assumed. Note that the field equations (10.2) corresponding to (11.4) are in general nonlinear (although still quasi-linear).

Example 11.2.5. Let η, $\overset{\circ}{\nabla}$, $\langle,\rangle$, f be as in Example 11.2.4, and let f assume its minimum value somewhere in $[0,\infty)$. By (11.5), the vacuum states for the Lagrangian (11.4) in η (as originally defined at the beginning of this section) are precisely those C^∞ sections ϕ of η with

$$\overset{\circ}{\nabla}\phi = 0, \qquad f(\langle\phi,\phi\rangle) = \min f, \tag{11.6}$$

provided that they exist. (Note that $\langle\phi,\phi\rangle$ is constant if $\overset{\circ}{\nabla}\phi = 0$.) As in Example 11.1.4, we may now drop the assumption (10.14) and still define the *vacuum states* for (11.4) to be those sections ϕ of η satisfying (11.6).

Example 11.2.6. The affine bundle $\mathcal{C}(\delta)$ (§5.3) for a vector bundle δ with a G-structure over $\mathcal{M}$ does not in general carry a "naturally distinguished" connection χ. However, a connection in $\mathcal{C}(\delta)$ can be provided by choosing any C^∞ *section* of $\mathcal{C}(\delta)$, i.e., a compatible connection ∇ in δ (usually interpreted as a vacuum state for an appropriate Lagrangian). In fact, ∇ induces the obvious isomorphism between $\mathcal{C}(\delta)$ and its translation-space bundle $\mathfrak{g}(\delta)T^* = \mathfrak{g}(\delta) \otimes T^*\mathcal{M}$, sending ∇ onto the zero section, while ∇ and the Levi-Civita connection of $(\mathcal{M}, g)$ clearly determine a (linear) connection in $\mathfrak{g}(\delta)T^*$.

Once such ∇ is chosen, any compatible connection in δ equals $\nabla + \Gamma$ for some section Γ of $\mathfrak{g}(\delta)T^*$, and we have the curvature formula

$$R^{\nabla+\Gamma} = R^\nabla - d^\nabla\Gamma - \Gamma \wedge \Gamma \tag{11.7}$$

(the details of which may vary, depending on notational conventions). The potential $\mathcal{V}$ of the Yang-Mills Lagrangian (10.6), corresponding to the connection χ in $\mathcal{C}(\delta)$ determined by ∇, is obtained from (10.6) and (11.7) by setting $\nabla\Gamma = 0$ (and hence $d^\nabla\Gamma = 0$), so that

$$\mathcal{V}(\nabla + \Gamma) = \frac{1}{4}\langle R^\nabla - \Gamma \wedge \Gamma, R^\nabla - \Gamma \wedge \Gamma\rangle. \tag{11.8}$$

Example 11.2.7. Let us interpret a vector bundle η, with $\overset{\circ}{\nabla}, \langle,\rangle, f$ and the Lagrangian (11.4) as in Example 11.2.5, as the *free-particle bundle* corresponding to some exotic kind of particles (governed by possibly nonlinear field equations), which are also subject to an interaction represented by the *interaction bundle* δ with a G-structure that includes a positive definite fibre metric in δ (cf. §5.4). The carriers of the interaction are as usual described by sections of $\mathcal{C}(\delta)$, i.e., compatible connections ∇ in δ, subject to the Yang-Mills equations (5.7), derived from the Lagrangian (10.6). The states ∇ of the carriers interact with those of the interacting matter particle, i.e., sections ψ of $\delta \otimes \eta$ (or, more generally, (5.12)) in a manner characterized by the general interaction Lagrangian (10.23), which in this case has the form

$$\mathcal{L}(\psi, \nabla) = -\frac{1}{2}\langle(\nabla \otimes \overset{\circ}{\nabla})\psi, (\nabla \otimes \overset{\circ}{\nabla})\psi\rangle - f(\langle\psi, \psi\rangle) - \frac{1}{4}\langle R^\nabla, R^\nabla\rangle, \tag{11.9}$$

the fibre metric in $\delta \otimes \eta$ (or (5.12)) being induced by those in δ and η. Assuming (10.14) and (10.16) we obtain from (11.1), (11.5) the corresponding energy-density formula $\mathcal{H}(\psi, \nabla) = \frac{1}{2}\langle(\nabla \otimes \overset{\circ}{\nabla})\psi, (\nabla \otimes \overset{\circ}{\nabla})\psi\rangle_+ + f(\langle\psi, \psi\rangle) + \frac{1}{4}\langle R^\nabla, R^\nabla\rangle_+$, so that the vacuum states (ψ, ∇) are characterized by

$$(\nabla \otimes \overset{\circ}{\nabla})\psi = 0, \qquad f(\langle\psi, \psi\rangle) = \min f, \qquad R^\nabla = 0, \tag{11.10}$$

cf. (11.6). As in Example 11.1.4, (11.10) can be used to define the *vacuum states* (ψ, ∇) *for the Lagrangian* (11.9), even without assuming (10.14) or (10.16).

Example 11.2.8. The affine bundle $(\delta \otimes \eta) \oplus \mathcal{C}(\delta)$ with the Lagrangian $\mathcal{L}$ given by (11.9) carries no natural connection χ. A "non-natural" connection χ can, however, be obtained by selecting a C^∞ section ∇ of $\mathcal{C}(\delta)$ and taking the direct sum of $\nabla \otimes \overset{\circ}{\nabla}$ in $\delta \otimes \eta$ and the connection determined by ∇ in $\mathcal{C}(\delta)$ (Example 11.2.6). Writing sections of $\mathcal{C}(\delta)$ in the form $\nabla + \Gamma$ (cf. Example 11.2.6) and formally setting $(\nabla \otimes \overset{\circ}{\nabla})\psi = 0$ and $\nabla\Gamma = 0$ (so that $d^\nabla\Gamma = 0$) in $\mathcal{L}(\psi, \nabla + \Gamma)$, we obtain the potential $\mathcal{V}$ of $\mathcal{L}$ (depending on the choice of ∇), with

$$\mathcal{V}(\psi, \nabla + \Gamma) = \frac{1}{2}\langle\Gamma\psi, \Gamma\psi\rangle + f(\langle\psi, \psi\rangle) + \frac{1}{4}\langle R^\nabla - \Gamma \wedge \Gamma, R^\nabla - \Gamma \wedge \Gamma\rangle. \tag{11.11}$$

This is immediate from (11.9) and the relation $(\nabla + \Gamma) \otimes \overset{\circ}{\nabla} = \nabla \otimes \overset{\circ}{\nabla} + \Gamma$, where Γ also stands for the bundle morphism $\Gamma \otimes \mathrm{Id}_\eta : \delta \otimes \eta \rightarrow \delta \otimes \eta \otimes T^*\mathcal{M}$.

References: Bleecker 1981; Ramond 1981; Leite Lopes 1981.

11.3 Vacua and Masses

Suppose that $\mathcal{V}$ is the potential of a Lagrangian $\mathcal{L}$ in a fibre bundle η, endowed with a fixed connection χ, over a spacetime $(\mathcal{M}, g)$ (§11.2). We say that a C^∞ section ψ of η is *critical* for $\mathcal{V}$ (or, *minimizes* $\mathcal{V}$) if the restriction of $\mathcal{V}$ to each fibre η_x of η, $x \in \mathcal{M}$, has a critical point (or, assumes its minimum value) at $\psi(x)$.

Let ζ be the *linearization* of η along a C^∞ section ψ, i.e., the real vector bundle $\zeta = \psi^* \upsilon$ over $\mathcal{M}$, where υ is the vertical subbundle of $T\eta$ (so that $\zeta_x = \upsilon_{\psi(x)} = T_{\psi(x)}\eta_x$ for $x \in \mathcal{M}$). A potential $\mathcal{V}$ in η for which ψ is critical then determines a C^∞ section $\tau = \mathrm{Hess}_\psi \mathcal{V}$ of $S^2 \zeta^*$, assigning to $x \in \mathcal{M}$ the Hessian τ_x of the restriction of $\mathcal{V}$ to η_x at the (critical) point $\psi(x)$. If, moreover, ψ minimizes $\mathcal{V}$, the symmetric bilinear form τ_x on ζ_x is positive semidefinite at each $x \in \mathcal{M}$.

We now assume that

(i) η is a fibre bundle over $\mathcal{M}$ with a connection χ, a Lagrangian $\mathcal{L}$, and the corresponding potential $\mathcal{V}$.

(ii) A given section ψ of η is a vacuum state for $\mathcal{L}$, as defined, in the appropriate cases, in Example 11.1.4 and Examples 11.2.5, 11.2.7 (without assuming (10.14)).

(iii) The linearization ζ of η along ψ carries a fixed, "natural" (real) fibre metric $\langle , \rangle_{\mathbb{R}}$ (which may, e.g., be induced by a natural fibre metric on the vertical subbundle υ of $T\eta$).

(iv) ψ is critical for $\mathcal{V}$. In the case where the fibre metric $\langle , \rangle_{\mathbb{R}}$ in ζ is positive definite, we also require ψ to minimize $\mathcal{V}$.

(v) Let F be the bundle endomorphism of ζ characterized by $\langle F_x u, v \rangle_{\mathbb{R}} = \tau_x(u, v)$ for $x \in \mathcal{M}$ and $u, v \in \zeta_x$. Our final assumptions are that the characteristic polynomial $\det(F_x - \mu \cdot \mathrm{Id}_{\zeta_x})$ of F does not depend on x and that each fibre ζ_x is an orthogonal direct sum of eigenspaces of F_x for nonnegative real eigenvalues. (Note that, whenever $\langle , \rangle_{\mathbb{R}}$ is positive definite, the latter condition follows from (iv), as F is self-adjoint.)

Under these hypotheses, we can use the fixed vacuum state ψ to *define the masses* of the particles represented by η. To this end, we let $\mu_1, \ldots, \mu_n$ be the mutually distinct, constant, nonnegative real eigenvalues of F in (v). The summands $\zeta_j = \mathrm{Ker}(F - \mu_j \cdot \mathrm{Id}_\zeta)$ of the $\langle , \rangle_{\mathbb{R}}$-orthogonal direct-sum decomposition

$$\zeta = \zeta_1 + \ldots + \zeta_n \qquad\qquad (11.12)$$

then are interpreted as models of particles with masses m_j characterized by $\mu_j = \frac{1}{2}(m_j c/\hbar)^{k_j}$, where $k_j \in \{1, 2\}$ has to be determined from additional considerations (cf. (11.2)). However, if a summand ζ_j admits a further natural decomposition, it usually describes several particle species of the same mass m_j (cf. §1.8). Thus, in general, a decomposition *finer* than (11.12) may be needed to represent all particles involved.

We now proceed to verify that in all models corresponding to one or several physical particles, discussed so far, conditions (i) – (v) are satisfied by *any* vacuum state ψ, and the mass values provided by the above definition agree with those obtained otherwise. (For more "exotic" cases, see the next section.)

Remark 11.3.1. If η is a vector bundle, its linearization ζ along any C^∞ section ψ is canonically isomorphic to η. Similarly, if η is an affine bundle, there is an obvious isomorphism betwen ζ and the translation-space bundle of η.

Remark 11.3.2. Although much information is usually lost when a fibre bundle η is replaced by its linearization along a C^∞ section, this will be of no consequence in our discussion, since we only deal with cases where η is a vector bundle, or an affine bundle (cf. Remark 11.3.1). In these cases, we may set $\zeta = \eta$ in (11.12) (or, in a suitable finer decomposition), and try to express $\mathcal{L}$ in the form (10.20), i.e., determine the Lagrangians $\mathcal{L}_j$ of the particles involved and the interaction term $\mathcal{L}_{\text{int}}$.

Remark 11.3.3. Despite the fact that the above definition of particle masses is based on fixing a vacuum state ψ, one may reasonably expect the mass values *not* to depend on the choice of ψ. This is for instance the case in all examples discussed below, often just because ψ is unique.

Example 11.3.4. Each of the Klein-Gordon, Dirac and Proca Lagrangians (10.3), (10.4), (10.11) in the appropriate vector bundle, with the potential (11.2) and the unique vacuum state $\psi = 0$ (Example 11.1.4(a)), satisfies (i) – (v) and leads to the expected mass value m, with a trivial (single-summand) decomposition (11.12).

Example 11.3.5. The Weyl Lagrangian (10.5) in a Weyl spinor bundle σ_{L} over $(\mathcal{M}, g)$ describes, according to the above definition (cf. Example 11.3.4), a single massless particle, provided that one chooses any real fibre metric $\langle , \rangle_{\mathbb{R}}$ in σ_{L} (see also §6.3).

Example 11.3.6. A direct sum involving several copies of vector bundles such as those in Example 11.3.4 (with arbitrarily prescribed masses) and

Example 11.3.5, along with the corresponding fibre metric, linear connection, and the interaction-free Lagrangian (10.26), has the unique vacuum state $\psi = 0$ (see Example 11.1.4(b)) and the potential $\mathcal{V}(\psi_1, \ldots, \psi_n) = \sum_j \mathcal{V}_j(\psi_j)$. Therefore (11.12), or its suitable finer version, coincides with the original decomposition, and the resulting masses are exactly the prescribed ones.

Example 11.3.7. Let ∇ be a fixed vacuum state for the Yang-Mills Lagrangian (10.6) in the affine bundle $\mathcal{C}(\delta)$, where δ is a vector bundle over $\mathcal{M}$ endowed with a G-structure and a natural fibre metric in $\mathfrak{g}(\delta)$. The corresponding potential (11.8) with $R^\nabla = 0$ (see Example 11.1.4(c)), restricted to each fibre, is a homogeneous quartic function of Γ, and hence its Hessian at $\Gamma = 0$ vanishes. The above discussion now leads to a trivial (one-summand) decomposition of $\mathcal{C}(\delta)$, or rather its translation-space bundle $\mathfrak{g}(\delta) \otimes T^*\mathcal{M}$, with the correct mass value $m = 0$ (see Remark 6.9.4).

Example 11.3.8. Applying the above definition of masses to the interaction Lagrangian (10.23) with η, δ as in Examples 11.3.6, 11.3.7, respectively, for a fixed vacuum state $(0, \nabla)$ with $R^\nabla = 0$ (see Example 11.1.4(d)), and the potential $\mathcal{V}$ satisfying an analogue of (11.11), we obtain the original mass values m_j for the matter fields in η and $m = 0$ for the Yang-Mills field. Moreover, (11.12) then is obtained by applying the appropriate decomposition of η to $(\delta \otimes \eta) \oplus \mathcal{C}(\delta)$, where $\mathcal{C}(\delta)$ is identified with $\mathfrak{g}(\delta) \otimes T^*\mathcal{M}$ via ∇.

References: Bleecker 1981; Leite Lopes 1981; Chaichian and Nelipa 1984; Bogolyubov and Shirkov 1983; Ramond 1981.

11.4 The Higgs Mechanism

Suppose that $\langle , \rangle$ is a positive definite fibre metric in a real or complex vector bundle η with a fixed connection and a Lagrangian $\mathcal{L}$, the potential $\mathcal{V}$ of which (§11.2) has the form (11.3) for some C^∞ function $f : [0, \infty) \to \mathbb{R}$. A section ϕ of η is easily seen to be critical for $\mathcal{V}$ (§11.3) if and only if

$$f'(\langle \phi, \phi \rangle)\phi = 0, \tag{11.13}$$

and then $\tau = \text{Hess}_\phi \mathcal{V}$ is given by

$$\tau = 2f'(\langle \phi, \phi \rangle)\langle , \rangle_{\mathbb{R}} + 2f''(\langle \phi, \phi \rangle)\langle \phi, \cdot \rangle_{\mathbb{R}} \otimes \langle \phi, \cdot \rangle_{\mathbb{R}}, \tag{11.14}$$

where $\langle , \rangle_{\mathbb{R}} = \text{Re}\langle , \rangle$ (so that the subscript $\mathbb{R}$ may be dropped in η is real). Now let us assume that ϕ is a vacuum state in η (cf. Example 11.2.5), critical for $\mathcal{V}$, and $\langle \phi, \phi \rangle = z_0$ for some real number $z_0 \geq 0$ such that $f(z_0) = \min f$.

Conditions (i) – (v) of §11.3 are thus satisfied and, in view of (11.13) and (11.14), the resulting decomposition (11.12) of $\zeta = \eta$ into the n summands ζ_j, along with the corresponding masses m_j, can be described as follows:

(a) If $z_0 f''(z_0) = 0$, then the last term in (11.14) vanishes, so that $n = 1$ and $\zeta_1 = \eta$, while $\mu_1 = \frac{1}{2}(m_1 c/\hbar)^{k_1} = 2f'(0)$ (hence, by (11.13), $m_1 = 0$ unless $z_0 = 0$).

(b) If $z_0 f''(z_0) \neq 0$ and the real fibre dimension of η is greater than one, we have $f'(\langle \phi, \phi \rangle) = 0$ (from (11.13)), which implies that $n = 2$ and $\eta = \zeta_1 + \zeta_2 = \mathbb{R}\phi + \phi^\perp$ (orthogonal complement relative to $\langle , \rangle_{\mathbb{R}}$), with

$$\mu_1 = \frac{1}{2}(m_1 c/\hbar)^{k_1} = 2z_0 f''(z_0) > 0 \qquad (11.15)$$

and $m_2 = 0$.

(c) Finally, in the case where $z_0 f''(z_0) \neq 0$ and η is a real line bundle, one sees as in (b) that $n = 1$, $\zeta_1 = \eta$ and the mass m_1 is given by (11.15).

The above discussion applies, in particular, to the Lagrangian (11.4) with η, $\overset{\circ}{\nabla}$, $\langle , \rangle$, f as in Examples 11.2.4, 11.2.5. Structures of this type, with $z_0 > 0$, can be used to make some Yang-Mills fields appear massive, as we will see next. However, interpreted as models of physical reality, they lead, in (a) and (b), to predictions of massless particles, known as the *Goldstone bosons*. In particular, if η is a product bundle (which is the simplest choice), the Goldstone bosons are both *massless* and *spinless* (§1.16, §1.8). Since no such objects seem to exist in nature, case (c) becomes the only plausible possibility.

Thus, let us consider a matter particle described by the Lagrangian (11.4) in the product bundle $\eta = \mathcal{M} \times \mathbb{R}$ over the spacetime $(\mathcal{M}, g)$ with the flat product connection $\overset{\circ}{\nabla} = d$ and the constant fibre metric $\langle , \rangle$, which is the ordinary multiplication of sections of η regarded as functions $\mathcal{M} \to \mathbb{R}$, and let f satisfy $f(z_0) = \min f$ for a *unique* $z_0 \geq 0$ that in addition has $z_0 f''(z_0) \neq 0$. Uniqueness of z_0 makes the mass m_1 of the particle, given by (11.15), independent of the choice of a vacuum state in η which minimizes $\mathcal{V}$, cf. Remark 11.3.3. Furthermore, let this particle be subject to an interaction described by an interaction bundle δ with some G-structure (including a positive definite fibre metric) and a natural fibre metric in $\mathfrak{g}(\delta)$ (§5.4, §6.8). The coupled field equations (5.9), (5.10) imposed on sections ψ of $\delta \otimes \eta = \delta$ and ∇ of $\mathcal{C}(\delta)$ then can be derived from the Lagrangian (11.9) in $\delta \oplus \mathcal{C}(\delta)$ (see Example 10.3.1) where, under the identifications $\delta \otimes \eta = \delta$ and $\overset{\circ}{\nabla} = d$, $\nabla \otimes \overset{\circ}{\nabla}$ may be replaced by ∇. Choosing a vacuum state (ϕ, ∇) in $\delta \oplus \mathcal{C}(\delta)$ as defined by (11.10), i.e., with

$$\nabla \phi = 0, \quad \langle \phi, \phi \rangle = z_0, \quad R^\nabla = 0, \qquad (11.16)$$

one easily sees that (ϕ, ∇) is critical for the corresponding potential $\mathcal{V}$ given by (11.11) with $R^\nabla = 0$ (§11.3). In view of (11.14), with $f'(z_0) = 0$,

$\mathrm{Hess}_{(\phi, \nabla)} \mathcal{V}$ regarded as a *quadratic* form in the linearization $\delta + \mathfrak{g}(\delta) T^* = \delta \oplus (\mathfrak{g}(\delta) \otimes T^* \mathcal{M})$ of $\delta \oplus \mathcal{C}(\delta)$ (cf. §11.3) is given by

$$(\xi, \varGamma) \mapsto \frac{1}{2} \langle \varGamma \phi, \varGamma \phi \rangle + 2 f''(z_0)(\mathrm{Re}\, \langle \phi, \xi \rangle)^2. \tag{11.17}$$

The corresponding decomposition of $\delta + \mathfrak{g}(\delta) T^*$, equal to or finer than (11.12) (and based on decomposing either of the summands δ, $\mathfrak{g}(\delta) T^*$), then may lead to *positive masses of some interaction carriers* (see §11.5 for specific examples), as the quadratic form $\varGamma \mapsto \frac{1}{2} \langle \varGamma \phi, \varGamma \phi \rangle$ in $\mathfrak{g}(\delta) T^*$ is usually nonzero. On the other hand, unless δ is a real line bundle (so that G has 1 or 2 elements, in contrast with the requirement that the physical gauge groups be infinite), the δ part of the decomposition and the resulting masses are exactly as described in (b) above, which leads, besides the original matter particle with the mass $m_1 > 0$, living in $\mathbb{R}\phi = \mathcal{M} \times \mathbb{R}$, also to some (massless) Goldstone bosons. The latter can, however, be disregarded by using an argument involving gauge symmetry (see Remark 11.4.1).

This idea of introducing an additional massive matter particle living in $\mathcal{M} \times \mathbb{R}$ and governed by a nonlinear field equation, and thus "generating" positive masses of some Yang-Mills fields, yet without producing any unwanted Goldstone bosons, is known as the *Higgs mechanism*.

Remark 11.4.1. For many physical purposes, two connections in an interaction bundle δ, compatible with its G-structure and obtained from each other by a C^∞ gauge transformation, may be regarded as "physically identical" (§5.9). A similar principle of *physical equivalence under gauge transformations* can, by analogy, be applied to sections of δ. By further stretching this idea, one may conclude that sections of the line bundle $\mathbb{R}\phi \subset \delta$ (with ϕ chosen as above) represent, essentially, all "physical" states in δ, since each section ϕ of δ becomes locally C^∞ gauge equivalent to a section of $\mathbb{R}\phi$ when restricted to a suitable open dense subset of $\mathcal{M}$.

Remark 11.4.2. Under the above assumptions on η, $\overset{\circ}{\nabla}$, $\langle, \rangle$, $\mathcal{L}$, f and z_0, we can rewrite (11.4) as

$$\mathcal{L}(\psi) = -\frac{1}{2} \langle d\psi, d\psi \rangle - f(\psi^2) \tag{11.18}$$

for functions $\psi : \mathcal{M} \to \mathbb{R}$. A fixed vacuum state ϕ in $\eta = \mathcal{M} \times \mathbb{R}$ then satisfies (11.6), i.e., is a constant real-valued function on $\mathcal{M}$ with $\phi^2 = z_0$. Thus, with $m = m_1$ given by (11.15) (where $k_1 = 2$),

$$\mathcal{L}(\phi + \xi) = -\frac{1}{2}(\langle d\xi, d\xi \rangle + \frac{m^2 c^2}{\hbar^2} \xi^2) - \min f + O(\xi^3),$$

which is obtained by summing the Klein-Gordon Lagrangian (10.3) with the correct mass value m, applied to $\xi : \mathcal{M} \to \mathbb{R}$, and some higher-order terms (as well as the constant $- \min f$). Nonlinearity of the resulting field equation (10.2) indicates that the particle in question is subject to nontrivial self-interaction (see end of §4.12), which in turn suggests (Remark 10.3.8) that the last term $O(\xi^3)$ should be polynomial in ξ, so that f is a polynomial. The simplest (lowest-order) choice of f that satisfies the above requirements concerning the existence and uniqueness of z_0 is

$$f(z) = az^2 - bz \qquad (11.19)$$

with constants $a, b > 0$. Then

$$z_0 = \frac{b}{2a}, \quad m = \frac{2\hbar}{c}\sqrt{b} \qquad (11.20)$$

where $m = m_1$ is given by (11.15) with $k_1 = 2$.

References: Leite Lopes 1981; Ryder 1985; Bogolyubov and Shirkov 1983; Huang 1982; Chaichian and Nelipa 1984; Sudbery 1986.

11.5 Dynamics of the Electroweak Model

The specific values of mass in each lepton generation (6.20), and the masses of the electroweak-interaction carriers (§6.5), can be explained in dynamical terms, i.e., derived from a suitable Lagrangian (§10.1). With that purpose in mind, one postulates the existence of a peculiar physical field permeating the empty space, called the *Higgs field*, the states of which are real-valued functions on the given spacetime $(\mathcal{M}, g)$ (that is, sections of the product line bundle $1_{\mathbb{R}} = \mathcal{M} \times \mathbb{R}$), subject to an appropriate field equation. According to §1.2, one may expect the Higgs field to manifest itself through some matter particles (its "quanta"), known as the *Higgs bosons*. Their basic properties, justified ex post by the correctness of the resulting lepton-mass and carrier-mass spectra, can be described as follows.

(a) The Higgs boson is represented by the free-particle bundle $1_{\mathbb{R}} = \mathcal{M} \times \mathbb{R}$, and obeys the *nonlinear* field equation

$$\Box\phi + 2b\phi - 4a\phi^3 = 0 \qquad (11.21)$$

with positive constants a, b, derived from the Lagrangian $\mathcal{L}$ given by (11.18) with (11.19), i.e.,

$$\mathcal{L}(\phi) = -\frac{1}{2}\langle d\phi, d\phi \rangle + b\phi^2 - a\phi^4. \qquad (11.22)$$

(These choices of the bundle and the function f in (11.19) are justified in the paragraph following (c) of §11.4 and in Remark 11.4.2.)

(b) The way Higgs bosons are involved in the electroweak interaction is based on the simplest possible scenario (§5.4), so that their interacting-particle bundle is

$$\iota \underset{\mathbb{R}}{\otimes} 1_{\mathbb{R}} = \iota,$$

where ι stands for the electroweak-interaction bundle (§6.5).

(c) In addition to the electroweak interaction, the Higgs bosons are also subject to a direct pointlike interaction with leptons (as in Remark 10.3.7), characterized by the Lagrangian term (11.26) below.

The properties just listed obviously set the Higgs bosons apart from ordinary matter particles (§§1.5, 1.16, 1.17, 5.4). Actually, (c) and nonlinearity of (11.21) make them appear rather like self-interacting (non-Abelian) Yang-Mills fields, cf. §5.4. Moreover, Higgs bosons have not been observed in nature, and so their physical status is unclear. See also Remark 11.5.3 below.

We now proceed to a more detailed discussion of the *electroweak model with the Higgs field*, using the same assumptions and notations (including ι, σ, σ_{L}, σ_{R}, α_{ew}, λ) as in §6.5. The *free-particle (generic-particle) bundle* is the direct sum $1_{\mathbb{R}} + \sigma$, where $1_{\mathbb{R}}$ represents the Higgs boson, and the fixed Dirac spinor bundle σ stands for a whole generation of leptons, which we for definiteness choose to be (e, ν_e). (See also Remark 11.5.1.) According to §1.8, (5.12), (b) and (6.21), the most obvious choice of the *interacting-particle bundle* is

$$\alpha = \iota + \alpha_{\mathrm{ew}} = \iota + \iota\alpha_{\mathrm{L}} + (\Lambda^2 \iota)\sigma_{\mathrm{R}}. \tag{11.23}$$

Carriers of the electroweak interaction are represented by the affine bundle $\mathcal{C}(\iota)$ (§6.5), and the manner in which they interact with the matter particles living in α can be described with the aid of a suitable Lagrangian $\mathcal{L}$ in the direct-sum affine bundle $\alpha + \mathcal{C}(\iota)$ (cf. §10.3). To define $\mathcal{L}$, it is convenient to identify C^∞ sections of $\alpha + \mathcal{C}(\iota)$ with triples (ϕ, ψ, ∇) consisting of sections of ι, α_{ew} and $\mathcal{C}(\iota)$, respectively. To such (ϕ, ψ, ∇), $\mathcal{L}$ should assign a real-valued function $\mathcal{L}(\phi, \psi, \nabla)$ on $\mathcal{M}$, given by a formula of type (10.20), where the $\mathcal{L}_j$ correspond to the Higgs, Dirac and Yang-Mills Lagrangians (11.22), (10.4), (10.6) (with some necessary adjustments, since ϕ, ψ do not live in $1_{\mathbb{R}}$, σ). It is also reasonable to require that the interaction term in (10.20) have the form (10.21), with the two-particle contributions involving ∇ both given by the minimum-coupling substitution (§10.3), which at the same time clarifies the "bundle adjustments" just mentioned. However, to account for the observed lepton-mass spectrum, (10.21) must also contain a term involving ϕ and ψ only, which amounts to a nontrivial, pointlike interaction between Higgs bosons and leptons (cf. (c) above). Specifically, one sets

$$\mathcal{L}(\phi, \psi, \nabla) = \mathcal{L}_{\mathrm{H}}(\phi, \nabla) + \mathcal{L}_{\mathrm{D}}(\psi, \nabla) + \mathcal{L}_{\mathrm{YM}}(\nabla) + \mathcal{L}_{\mathrm{point}}(\phi, \psi) \tag{11.24}$$

with

$$\mathcal{L}_{\mathrm{H}}(\phi, \nabla) = -\frac{1}{2}\langle \nabla\phi, \nabla\phi \rangle + b\langle \phi, \phi \rangle - a\langle \phi, \phi \rangle^2,$$

$$\mathcal{L}_{\mathrm{D}}(\psi, \nabla) = \frac{1}{2}\mathrm{Im}\,\omega(\psi, \mathcal{D}^{\nabla}\bar{\psi}), \qquad (11.25)$$

$$\mathcal{L}_{\mathrm{YM}}(\nabla) = -\frac{1}{4}\langle R^{\nabla}, R^{\nabla} \rangle,$$

$$\mathcal{L}_{\mathrm{point}}(\phi, \psi) = -C_e \mathrm{Im}\{\psi_{\mathrm{L}}, \phi, \bar{\psi}_{\mathrm{R}}\}, \qquad (11.26)$$

where the following notations are used. In (11.26), C_e is a positive constant characterizing the strength (§6.8) of the lepton-Higgs boson pointlike interaction, while ω and $\mathcal{D}^{\nabla}$ in (11.25) are defined in §1.3 (or §6.3) and Remark 5.4.2. Finally, $\{\ ,\ ,\ \}$ in (11.26) stands for the trilinear map corresponding to the bundle morphism $\iota\sigma_{\mathrm{L}}\iota(\varLambda^2\bar\iota)\bar\sigma_{\mathrm{R}} = \iota\iota(\varLambda^2\iota^*)\sigma_{\mathrm{L}}\sigma_{\mathrm{L}} \to 1$ (notation of §5.1), which is the tensor product of $\omega : \sigma_{\mathrm{L}}\sigma_{\mathrm{L}} \to 1$ and the evaluation map $\iota\iota(\varLambda^2\iota^*) \to 1$, with $\bar\iota = \iota^*$ and $\bar\sigma_{\mathrm{R}} = \sigma_{\mathrm{L}}$ as in (vi) of §7.3 and (6.3), while the sections $\psi_{\mathrm{L}}, \psi_{\mathrm{R}}$ of $\iota\sigma_{\mathrm{L}}$, $(\varLambda^2\iota)\sigma_{\mathrm{R}}$ are the components of ψ under the decomposition (6.21) of α_{ew}.

Note that (11.25) corresponds to (10.4) with $m = 0$, so that generic leptons are "initially" regarded as massless, and it is only through symmetry breaking (see below) that some of them acquire positive masses. Also, the choice of (11.26) is consistent with Remark 10.3.8.

Combining Example 11.1.4(b) with (11.16) and (11.20), it is now natural to define vacuum states for the Lagrangian (11.24) to be triples $(\phi, 0, \nabla)$ such that ∇ is a flat U(2)-connection in ι, ϕ is a section of ι with $\nabla\phi = 0$ and $2a\langle \phi, \phi \rangle = b$, and 0 stands for the zero section of α_{ew}. Thus, the kind of symmetry breaking based on fixing ϕ with the stated properties, used in §6.5, may be thought of as a *consequence* of selecting a vacuum state $(\phi, 0, \nabla)$ in $\alpha + \mathcal{C}(\iota)$. Moreover, the fixed flat connection ∇ in ι, along with the spinor connection $\overset{\circ}{\nabla}$ in σ, then determines a connection in $\alpha + \mathcal{C}(\iota)$ (Example 11.2.8) and the resulting potential for (11.24) is

$$\mathcal{V}(\phi + \xi, \psi, \nabla + \varGamma) = \frac{1}{2}\langle \varGamma(\phi + \xi), \varGamma(\phi + \xi) \rangle + a\langle \phi + \xi, \phi + \xi \rangle^2$$

$$-b\langle \phi + \xi, \phi + \xi \rangle + \frac{1}{4}\langle \varGamma \wedge \varGamma, \varGamma \wedge \varGamma \rangle + C_e\mathrm{Im}\{\psi_{\mathrm{L}}, \phi + \xi, \bar\psi_{\mathrm{R}}\}$$

(cf. (11.11)). In fact, one obtains $\mathcal{V}$ as in §11.2, replacing (ϕ, ψ, ∇) in (11.24) – (11.26) by $(\phi+\xi, \psi, \nabla+\varGamma)$ (since ϕ, ∇ now are fixed), and formally setting $\nabla\xi, \mathcal{D}^{\nabla}\psi, d^{\nabla}\varGamma$ equal to zero (while $\nabla\phi = 0$, $R^{\nabla} = 0$ by assumption). It is now immediate that the vacuum state $(\phi, 0, \nabla)$ is critical for $\mathcal{V}$ and leads to $\tau = \mathrm{Hess}_{(\phi,0,\nabla)}\mathcal{V}$ (§11.3) the quadratic form of which is

$$(\xi, \psi, \varGamma) \mapsto 4a(\mathrm{Re}\langle \phi, \xi \rangle)^2 + C_e\mathrm{Im}\{\psi_{\mathrm{L}}, \phi, \bar\psi_{\mathrm{R}}\} + \frac{1}{2}\langle \varGamma\phi, \varGamma\phi \rangle, \qquad (11.27)$$

cf. (11.17).

With the aid of the fixed section $(\phi, 0, \nabla)$, the affine bundle $\alpha + \mathcal{C}(\iota)$ is identified with its linearization (translation-space bundle) $\zeta = \alpha + \mathfrak{u}(\iota)T^*$. As in §11.3, (11.27) now leads to a natural decomposition

$$\alpha + \mathcal{C}(\iota) = \zeta = 1_\mathbb{R} + 1_\mathbb{R} + \lambda + \sigma_\mathrm{L} + \lambda\sigma + T^* + \lambda T^* + T^* \qquad (11.28)$$

finer than (11.12), with the mass values

$$m_\mathrm{H}, 0, 0, m_\nu = 0, m_\mathrm{e}, m_\gamma = 0, m_\mathrm{W}, m_\mathrm{Z}, \qquad (11.29)$$

corresponding to the eight summands of (11.28), and given by

$$m_\mathrm{H} = \frac{2\hbar}{c}\sqrt{b}, \quad m_\mathrm{e} = \frac{\hbar}{c}C_\mathrm{e}\frac{b}{a}, \quad m_\mathrm{W} = \frac{\hbar}{c}\sqrt{\frac{b}{8p_0 a}}, \quad m_\mathrm{Z} = \frac{1}{\cos\theta}m_\mathrm{W}, (11.30)$$

where a, b, C_e are the positive constants appearing in (11.22), (11.26), $p_0 > 0$ and the Weinberg angle θ are ingredients of the geometry of ι introduced in §6.5, while $\hbar, c$ as always denote Planck's constant divided by 2π and the speed of light. Relations (11.28) – (11.30) are immediate from (11.27) along with $1 = 1_\mathbb{R} + 1_\mathbb{R}$, (6.23), (6.24), and (6.30) with $\mathcal{C}(\lambda) = T^*$ (as established by the projection $\nabla^{[\lambda]}$ of ∇ onto λ, cf. (ix) of §6.6), as well as (6.7) and (6.18).

The second and third summands $1_\mathbb{R}, \lambda$ in (11.28) are believed *not* to represent any physical particles, since one can use local gauge transformations of $\iota = 1_\mathbb{R} + 1_\mathbb{R} + \lambda$ to transform their sections, almost everywhere, into sections of the first summand $1_\mathbb{R}$. (See Remarks 11.4.1, 11.5.4.) The remaining six summands $1_\mathbb{R}, \sigma_\mathrm{L}, \lambda\sigma, T^*, \lambda T^*, T^*$ are interpreted (in agreement with §§1.8, 1.16, 1.17, 6.5) as models of the six particle species "emerging" from the symmetry breaking: the Higgs boson, electronic neutrino ν_e, electron e, photon γ, W boson and Z boson, respectively, with masses given by (11.29) – (11.30). (About their antiparticles, see (iv) of §6.6.) The resulting theory, involving $a, b, C_\mathrm{e}, p_0, \theta$ as parameters, correctly predicts zero mass for ν_e, γ and positive masses $m_\mathrm{H}, m_\mathrm{e}, m_\mathrm{W}, m_\mathrm{Z}$ of the other particles, the last two being obtained with the aid of the Higgs mechanism (§11.4). Moreover, measurements of $m_\mathrm{e}, m_\mathrm{W}, m_\mathrm{Z}$ and the electron charge q, combined with (6.37), (11.30), lead to unique *experimental values* of $C_\mathrm{e}, p_0, \theta$ and b/a, in a manner consistent with numerous further observations. At the same time, $b > 0$ (and hence also m_H) remains a free parameter of the theory. Thus, the lack of evidence for the Higgs bosons in nature can be attributed to m_H still lying beyond the energy range of available colliders. (See Remark 6.9.5.)

The decomposition (11.28) not only leads to correct mass values for the particles involved, but also provides a plausible interpretation of (11.24) as an interaction Lagrangian. Specifically, replacing (ϕ, ψ, ∇) in (11.24) by $(\phi + \xi, \psi, \nabla + \Gamma)$ (with ϕ, ∇ fixed as above), decomposing ξ, ψ, Γ according to (11.28), and ignoring the "unphysical" summands $1_{\mathbb{R}}$, λ, one obtains an expression of type (10.20). The $\mathcal{L}_j$ then are the expected free-particle Lagrangians, given by (11.22) for the Higgs boson, (10.5) for ν_{e}, (10.4) with $m = m_{\mathrm{e}}$ for the electron e, and by (10.11) with m equal to 0, m_{W} or m_{Z} for the photon γ and W, Z bosons, respectively. (Cf. Remark 10.1.7.) The interaction terms in (10.20) are all polynomial, in agreement with Remark 10.3.8, and their particular form is consistent with available physical data. For instance, there is no interaction term involving ν_{e} and γ, which accounts for the well-known fact that neutrinos do not interact electromagnetically.

For the physical significance of choosing a vacuum state, see Remark 11.5.2 below.

Remark 11.5.1. To adapt the above model to *any* of the lepton generations (6.20), it suffices to replace C_{e} in (11.26) by a more general positive constant (C_{e}, C_{μ} or C_{τ}). This leaves all masses (11.29), (11.30) unchanged except for the electron mass m_{e}, which may have to be replaced by the muon mass m_{μ} or tauon mass m_{τ}. As a consequence, the lepton-Higgs boson coupling constant is generation-dependent, while the other parameters (a, b, p_0, θ) are not.

Remark 11.5.2. The symmetry breaking in the electroweak model, believed to actually take place in nature (§6.5), appeared in the above discussion as a by-product of fixing a vacuum state ($\phi, 0, \nabla$). On the other hand, one often chooses vacuum states with the sole purpose of defining particle masses (§11.3), which in turn should not depend on the vacuum used (Remark 11.3.3). The presence of one preferred vacuum may in fact be too restrictive for the model in question, enriching its geometry beyond what physics calls for. This is why it seems reasonable to expect that nature only singles out the ϕ part of an electroweak vacuum ($\phi, 0, \nabla$), which in itself is just a section ϕ of ι with $2a\langle\phi, \phi\rangle = b$ (cf. §6.5). Without a fixed ∇, the first T^* summand in (11.28) has to be replaced back by $\mathcal{C}(\lambda)$ (as in (6.30)), while the free-photon term of the corresponding decomposition (10.20) of the Lagrangian (11.24) then is given by (10.6) with $\nabla = \nabla^{[\lambda]}$. As a result, the massless photon still appears as a U(1) gauge field, while its massive colleagues W and Z are regarded as matter fields, in agreement with §6.9 and the second paragraph of §6.5.

Remark 11.5.3. One usually expects that the Higgs boson will turn out to be a matter particle, albeit very heavy and displaying some peculiar properties (see the paragraph following (c) in this section, and Remark 11.5.4 below). However, it is also conceivable that the Higgs mechanism in the electroweak model could be given a different physical interpretation, not invoking a hypothetical particle.

Remark 11.5.4. A further difference between Higgs bosons and ordinary matter particles is that, whereas the former enjoy full $U(2)$ gauge symmetry (which leads to elimination of Goldstone bosons, as in §11.4, and earlier in this section), the latter only do that for the reduced gauge group (in this case, $U(1)$), obtained from symmetry breaking. In fact, the distinction between neutrinos and massive leptons only comes into existence when the symmetry is broken.

References: Ryder 1985; Leite Lopes 1981; Huang 1982; Bogolyubov and Shirkov 1983.

References

Besse, A.L. (1987): *Einstein Manifolds*, Ergebnisse, ser. 3, vol. 10 (Springer-Verlag, Berlin)

Binz, E., Śniatycki, J., Fischer, H. (1988): *Geometry of Classical Fields*, North-Holland Mathematics Studies, Vol. 154 (North-Holland, Amsterdam)

Birrell, N.D., Davies, P.C.W. (1983): *Quantum Fields in Curved Space* (Cambridge Univ. Press, Cambridge)

Bleecker, D. (1981): *Gauge Theory and Variational Principles*, Global Analysis, Pure and Applied, Series A, No. 1 (Addison-Wesley, Reading, Mass.)

Bogolyubov, N.N. [Bogolubov, N.N.], Logunov, A.A., Oksak A.I., Todorov, I.T. (1990): *General Principles of Quantum Field Theory*, Mathematical Physics and Applied Mathematics, Vol. 10 (Kluwer, Dordrecht)

Bogolyubov, N.N. [Bogoliubov, N.N.], Shirkov, D.V. (1983): *Quantum Fields* (Benjamin/Cummings, Reading, Mass.)

Chaichian, M., Nelipa, N.F. (1984) *Introduction to Gauge Field Theories*, Texts and Monographs in Physics (Springer-Verlag, Berlin)

Derdzinski, A. (1991a): "Geometry of Elementary Particles" (to appear)

Derdzinski, A. (1991b): *Notes on Particle Physics* (Ohio State Math. Research Institute preprints)

DeWitt, B.S. (1965): *Dynamical Theory of Groups and Fields* (Gordon and Breach, New York)

Dixon, W.G. (1978): *Special Relativity: The Foundation of Macroscopic Physics* (Cambridge Univ. Press, Cambridge)

Emch, G.G. (1984): *Mathematical and Conceptual Foundations of 20th-Century Physics*, North-Holland Mathematics Studies, Vol. 100 (North-Holland, Amsterdam)

Faddeev, L.D., Slavnov, A.A. (1980): *Gauge Fields, Introduction to Quantum Theory* (Benjamin/Cummings, Reading, Mass.)

Felsager, B. (1983): *Geometry, Particles and Fields* (Odense Univ. Press, Odense)

Friedlander, F.G. (1975): *The Wave Equation on a Curved Space-Time* (Cambridge Univ. Press, Cambridge)

Fritzsch, H., Minkowski, P. (1975): "Unified interactions of leptons and hadrons", Ann. Phys., Vol. 93, pp. 193–266

Griffiths, D.J. (1989): *Introduction to Electrodynamics* (Prentice Hall, Englewood Cliffs, N.J.)

Hayward, R,W. (1976): *The Dynamics of Fields of Higher Spin*, National Bureau of Standards Monograph 154 (U. S. Government, Washington, D.C.)

Hermann, R. (1970): *Vector Bundles in Mathematical Physics* , Vols. I,II (Benjamin, New York)

Hermann, R. (1973–1988): *Interdisciplinary Mathematics, Vols. I–XXIV* (Math Sci Press, Brookline, Mass.)

Huang, K. (1982): *Quarks, Leptons and Gauge Fields* (World Scientific, Singapore)

Kobayashi, S., Nomizu, K. (1963): *Foundations of Differential Geometry, Vol. I* (Interscience, New York)

Lawson, H.B., Michelsohn, M.-L. (1989): *Spin Geometry*, Princeton Math. Series, Vol. 38 (Princeton Univ. Press, Princeton)

Leite Lopes, J. (1981): *Gauge Field Theories* (Pergamon Press, Oxford)

Lichtenberg, D.B. (1978): *Unitary Symmetry and Elementary Particles*, 2nd ed. (Academic Press, New York)

Mackey, G.W. (1963): *The Mathematical Foundations of Quantum Mechanics* (Benjamin, New York)

Manin, Y.I. (1981): *Mathematics and Physics*, Progress in Physics, Vol. 3 (Birkhäuser, Boston)

Manin, Y.I. (1988): *Gauge Field Theory and Complex Geometry* (Springer-Verlag, Berlin)

Milnor, J.W., Stasheff, J.D. (1974): *Characteristic Classes*, Annals of Mathematics Studies, Vol. 76 (Princeton Univ. Press, Princeton)

Mohapatra, R.N. (1986): *Unification and Supersymmetry*, Contemporary Physics (Springer-Verlag, New York)

Palais, R.S., Terng, C.-L. (1977): "Natural Bundles Have Finite Order", Topology, Vol. 16, pp. 271–277

Parrott, S. (1987): *Relativistic Electrodynamics and Differential Geometry* (Springer-Verlag, New York)

Penrose R., Rindler, W. (1984–1986): *Spinors and Space-Time I, II* (Cambridge Univ. Press, Cambridge)

Ramond, P. (1981): *Field Theory* (Benjamin/Cummings, Reading, Mass.)

Roman, P. (1961): *Theory of Elementary Particles* (North-Holland, Amsterdam)

Ryder, L.H. (1986): *Elementary Particles and Symmetries*, Documents on Modern Physics (Gordon and Breach, New York)

Ryder, L.H. (1985): *Quantum Field Theory* (Cambridge Univ. Press, Cambridge)

Sachs, K., Wu, H. (1977): *General Relativity for Mathematicians* (Springer-Verlag, New York)

Schweber, S.S. (1962): *An Introduction to Relativistic Quantum Field Theory* (Harper & Row, New York)

Sudbery, A. (1986): *Quantum Mechanics and the Particles of Nature: An Outline for Mathematicians* (Cambridge University Press, Cambridge)

Thirring, W. (1978–1983): *A Course in Mathematical Physics, Vols. 1-4* (Springer-Verlag, New York)

Zhelobenko, D.P. [Želobenko, D.P.] (1973): *Compact Lie Groups and Their Representations* (AMS Translations, Providence, R.I.)

Subject Index

196   Subject Index

Electroweak
  interaction   83, 106 – 112
  model, geometry   107 – 111
  model, dynamics   186 – 191
  model, for quarks   161 – 163
Electroweak-interaction bundle   107,
123
Elementary charge   73
Energy density   169 – 171
Equation of continuity   64, 65
Euler-Lagrange equations   31, 69, 164
Evolving state   4, 20

Fermion   11, 14
Fermionic representation   26, 52, 56
Field equations   4, 164
  coupled   84
  free   76, 84
  linear   77
Field quantization   113
First-order natural bundle   20
Flavor   9, 98
  symmetries   138 – 148
Four-current   65
Four-momentum   66
Fractional charges   99
Free Lagrangian   69
Free particle   17
Free-particle bundle   76, 84
Future-pointing timelike vector   3

G-parity   41, 130
  conservation of   41
Gauge
  equivalence   92
  field   7, 80, 84
  hierarchies   128
  transformation   82
Gell-Mann–Nishijima relation   42, 131,
132
General theory of relativity   3
Generalizations of particles   9
Generalized
  charges   8, 41
  hadrons   138 – 139
  momentum components   31
  position components   31
  spin bundle   9
  spin symmetry   9
Generic-fermion bundle   123, 161
Generic-lepton bundle   107
Glashow-Illiopoulos-Maiani model   155,
161 – 163
Glashow-Weinberg angle   109
Glueballs   101
Gluons   6, 43, 100
Goldstone bosons   176, 184, 191
Grand unified theories   123, 127 –
128
Gravitational interaction   5 – 6

Gravitons   6
Gravity   3, 6

Hadrons   7, 11, 14 – 16, 98 – 100,
138 – 154
Hamiltonian   69
Hamilton's equations   31, 72
Harmonic bundle   94
Helicity   45, 130
Hermitian plane   104
Higgs
  boson   186
  field   186
  mechanism   185, 189
Hypercharge   37, 42, 130 – 132, 142

Indeterminism   31
Indistinguishability of particles   11
Interacting-fermion bundle   123
Interacting-lepton bundle   107
Interacting-particle bundle   76, 84
Interaction   5
  bundle   83
  carriers   6, 12, 84, 88, 91
  charge   88
  Lagrangian   70, 171 – 173
  range   106, 111, 119
  strength   5, 115
Interatomic forces   6
Intermediate bosons   6
Intermolecular forces   6
Internal energy   66
Interquark forces   6
Isochronous Lorentz group   49
Isospin   9, 36, 130, 148 – 149
  class   45, 130
  component   36, 130, 141, 142
  conservation law   44
  geometry   37 – 38
  I bundle   37
  in nature   36 – 37
Isospinor bundle   37

Kinetic energy   66
Kinetic term   178
Klein-Gordon equation   17 – 19, 59,
68, 96, 165
Kobayashi-Maskawa
  matrix   158
  model   159 – 160
  parameters   158, 160

Lagrangian   69
  density   164
Lambda hyperon   14
Lepton   7, 11, 12, 14
  generations   107
  masses   122
  number   41, 130
  proper   14, 42
Lichnerowicz formula   18